Identification of electrochemical reaction kinetics by dynamic methods

I0131835

Identification of electrochemical reaction kinetics by dynamic methods

Von der Fakultät für Maschinenbau
der Technischen Universität Carolo-Wilhelmina zu Braunschweig

zur Erlangung der Würde

eines Doktor-Ingenieurs (Dr.-Ing.)

genehmigte Dissertation

von: Dipl.-Wirtsch.-Ing. Fabian Kubannek
aus (Geburtsort): Twistringen

eingereicht am: 27.03.2019
mündliche Prüfung am: 18.06.2019

Gutachter:

Frau Prof. Dr.-Ing. Ulrike Krewer
Herr Prof. Dr. Uwe Schröder

2019

Bibliografische Information der Deutschen Nationalbibliothek
Die Deutsche Nationalbibliothek verzeichnet diese Publikation in der Deutschen
Nationalbibliografie; detaillierte bibliografische Daten sind im Internet über
http://dnb.d-nb.de abrufbar.
1. Aufl. - Göttingen: Cuvillier, 2019
Zugl.: (TU) Braunschweig, Univ., Diss., 2019

© CUVILLIER VERLAG, Göttingen 2019
Nonnenstieg 8, 37075 Göttingen
Telefon: 0551-54724-0
Telefax: 0551-54724-21
www.cuvillier.de

ISBN 978-3-7369-7054-0

Abstract

Electrochemical energy conversion technologies such as fuel cells, electrolysers, and batteries are often seen as key components for the transition to an economy that is powered by renewable energy sources. Knowledge-based design and systematic improvement of electrochemical processes will only become possible in future if the underlying reaction kinetics are well-understood.

The combination of dynamic electrochemical methods and dynamic measurement of bulk concentration and surface species led to the development of various *in operando* techniques that have been shown to yield additional insights into electrode processes. However, usually not all state variables and quantities of interest in an electrochemical system can be accessed experimentally even with advanced analytical techniques. Modelling and simulations are suitable tools to investigate mechanisms, extract kinetic parameters and observe state variables that cannot be measured directly in electrochemical systems.

While dynamic electrochemical methods are frequently coupled with *in operando* analytical techniques or with model-based analysis of electrochemical systems, the combination of all three approaches is not very common. This thesis is based on the hypothesis that a combination of dynamic electrochemical methods, *in operando* techniques for the detection of chemical species, and simulations, is a feasible and advantageous way towards the determination of electrochemical reaction kinetics.

In order to demonstrate the feasibility and advantages of such a combined approach, four model systems are studied in this thesis.

In the first part, differential electrochemical mass spectrometry data and electrochemical data is used to parameterise physical models of the CO and methanol electrooxidation. The oxidation of CO on a porous Pt/Ru catalyst in this cell is analysed using potential step experiments. The characteristic responses of current density and CO_2 production rate allow to identify rate constants and transport coefficients using just one set of experimental data. The methodology is extended towards the more complex methanol electrooxidation reaction. With mass spectrometric flux-based frequency response analysis a new dynamic technique is introduced that allows to evaluate the production of volatile species as a function of the electrical current in the frequency domain. The prospects and limitation of the technique and the interpretation of the spectra with the help of a mathematical model is demonstrated.

The second part of the thesis covers bioelectrochemical reactions that feature a higher degree of complexity and a lack of mechanistic data that can be used as a foundation for kinetic analysis.

The first DEMS results on acetate oxidation in electrochemically active biofilms are presented, and storage mechanisms for charge as well as substrate are quantified. Furthermore, conversion pathways and rate constants in bioelectrochemical glycerol oxidation are investigated by applying concentration pulses of the reaction intermediates. The dynamic responses of current density and concentration allow to study the role of the intermediates in the overall reaction and to extract conversion rates of each intermediate. Based on the experiments, a model for a biofilm electrode that describes dynamics of biofilm growth and acetate oxidation is set up and parameterised rigorously.

In conclusion, it is demonstrated that the identification of electrochemical macro- and microkinetics can benefit significantly from the following factors: the application of dynamic techniques such as potential steps, cyclic voltammograms, and electrochemical impedance spectroscopy because they allow to separate processes by their time constants; additional concentration measurements apart from current and potential because transport processes and chemical reaction steps might only have an indirect influence on current and potential; physical simulation models that includes mass transfer and - if applicable - concentration and potential gradients over the electrode because they allow to access parameters that cannot be deduced directly from experimental data; a detailed analysis of the electrochemical cell that might include computational fluid dynamics analysis because this information is often necessary to evaluate mass transfer effects.

This thesis provides new physical insight into the analysed example systems, and the developed methodologies constitute a toolbox for dynamic analysis of kinetics in electrochemical systems.

Acknowledgement

This thesis was conducted at the Institute of Energy and Process Systems Engineering at Technische Universität Braunschweig. I am very grateful to my supervisor Prof. Ulrike Krewer for her valuable advice, support, and guidance in the last five years. She put a lot of effort into guiding me to better scientific results. Beyond science, she also was a mentor and role model for my personal development.

I also thank Prof. Uwe Schröder for being the second referee for my thesis and for fruitful discussions on bioelectrochemical systems, and Prof. Antje Spieß for being the head of the board of examiners.

Furthermore, I want to express my deep gratitude to my family for a lot of love and support throughout my education. I also thank my beloved wife Sandra for being patient with me working on my thesis when I should have spent time with her, for proofreading various versions of the manuscript, and for being interested in my research work.

I thank all my current and former colleagues at InES for the great working atmosphere and inspiring discussions about research and many other topics. It has been a pleasure working with you all!

I would also like to acknowledge my former students Dan Ye, Hengdi Li, Bo Yuan, Barbara Dziobek, Jonas Perk, Joana Kühne and Hans Klesser for supporting parts of the experimental work.

Finally I thank Tanja Vidaković-Koch for fruitful discussions about the cyclone cell, the group of Helmut Baltruschat in Bonn for valuable advice on conducting DEMS experiments, Waldemar Sauter for performing HPLC measurements for the glycerol experiments and Christopher Moß for helping to set up the glycerol experiments.

Contents

List of Symbols and Abbreviations

Symbols that are solely used locally may not be found in this list but are explained at the place where they are used.

Abbreviations

AC	Alternating Current
ARB	Anode Respiring Bacteria
BES	Bioelectrochemical System
CA	Chronoamperommetry
CFD	Computational Fluid Dynamics
cFRA	Concentration-alternating Frequency Response Analysis
COD	Chemical Oxygen Demand
CP	Chronopotentiometry
CV	Cyclic Voltammetry
DC	Direct Current
DEMS	Differential Electrochemical Mass Spectrometry
DMFC	Direct Methanol Fuel Cell
EAB	Electroactive Bacteria
EET	Extra-cellular Electron Transfer
EIS	Electrochemical Impedance Spectroscopy
EPIS	Electrochemical Pressure Impedance Spectroscopy
EPS	Extra-cellular Polymeric Substance
EQCM	Electrochemical Quartz Crystal Microbalance
FRA	Frequency Response Analysis
FTIRS	Fourier-Transform Infrared Spectroscopy
GDL	Gas diffusion layer
HER	Hydrogen Evolution Reaction
HPLC	High Performance Liquid Chromatography
MEA	Membrane Electrode Assembly
MEC	Microbial Electrolysis Cell
MFC	Microbial Fuel Cell
MIMS	Membrane Inlet Mass Spectrometry

MOR	Methanol Oxidation Reaction
MSCV	Mass Spectrometric Cyclic Voltammetry
MS	Mass Spectrometer
NFRA	Non-linear Frequency Response Analysis
OCV	Open Circuit Voltage
ORR	Oxygen Reduction Reaction
PEMFC	Polymer Electrolyte Membrane Fuel Cell
RDE	Rotating Disk Electrode
SEM	Scanning Electron Microscopy
SHE	Standard Hydrogen Electrode
SOFC	Solid Oxide Fuel Cell

Greek Symbols

$\alpha_{OH/CO}$	charge transfer coefficient
$\beta_{OH/CO}$	symmetry factor
δ	layer thickness (m)
δ_{bf}	biofilm thickness (m)
δ_v	hydrodynamic boundary layer thickness (m)
ϵ	porosity
η	reaction overpotential (V)
μ, ν	singular vector
ρ_{bf}	biofilm density (kg/m^3)
σ	standard deviation of measurements (A/m^2 or mol/m^3)
θ	parameter vector
θ_{CO}	relative surface coverage of CO
θ_{OH}	relative surface coverage of OH
ν	kinematic viscosity (m^2/s)
ω	angular velocity / frequency (1/s)
ζ	singular value

Subscripts

ac	acetate
ad	adsorbed
bf	biofilm
CH_3OH	methanol
CO_2	carbon dioxide
CO	carbon monoxide
conc	concentration
el	electrode

i	running index
i	electric current
j	running index
k	running index
MS	mass spectrometer
n	normalised value
OH	hydroxide

Superscripts

A	anode compartment
AC	anode catalyst layer
DL	diffusion layer
el	electrode
exp	experiment
in	inflow
M	membrane
out	outflow
sim	simulation

Latin Symbols

A_{el}	electrode area (m^2)
c	concentration (mol/m^3)
C_{dl}	capacity of the double layer (F)
CE	coulomb efficiency
c_∞	bulk concentration (mol/m^3)
c_{Pt}	number of platinum surface sites (mol/m^2)
c_{Ru}	number of ruthenium surface sites (mol/m^2)
D	diffusion coefficient (m^2/s)
d_{in}	diameter of the inlet tube (m)
E	electrode potential (V)
E^0	equilibrium potential (V)
E_{dl}	potential over the double layer (V)
$E_{external}$	uncorrected electrode potential including ohmic drop (V)
E_{KA}	half-saturation potential (V)
f_0	biomass productivity factor
F	Faraday constant ($= 96485\,\text{C/mol}$)
$g_{OH/CO}$	interaction factor for OH/CO
G_{MS}	flux-based frequency response analysis transfer function
I	total current (A)

i	current density (A/m^2)
i_{cell}	external cell current density (A/m^2)
i_{lim}	limiting current density (A/m^2)
I_{MS}	ion current (A)
$I_{reaction}$	current from electrochemical reactions (A)
K	DEMS calibration constant (C/mol)
k	reaction rate constant (1/s)
K^*	MS calibration constant (C/mol)
$k_{CO,ad/ox}$	adsorption / oxidation rate constant (1/s)
k_d	biomass decay factor (1/s)
$k_{OH,ad/de}$	adsorption / desorption rate constant (1/s)
K_S	half-saturation rate constant (mol/m^3)
$K_{S,0}$	minimum half-saturation rate constant at a biofilm thickness of zero (mol/m^3)
$K_{S,eff}$	effective half-saturation rate constant of a biofilm (mol/m^3)
M_{bf}	molar mass of the biofilm biomass per carbon atom (kg/mol)
N	collection efficiency of the DEMS cell
n	total number of data points
N_A	Avogadro constant (=6.022·10^{13} 1/mol)
$\dot{n}$	molar flux (mol/s)
$\dot{n}_I$	maximum molar flux into the vacuum (mol/s)
$\dot{n}_{MS}$	molar flux into the vacuum (mol/s)
$N_{surface}$	number of surface sites available for adsorption
p	pressure (Pa)
Q_{dl}	charge in the double layer (C)
Q_m	charge for one monolayer of monovalent adsorbate (C)
q_{max}	maximum specific substrate consumption rate (mol/s/kg)
r	reaction rate (mol/m^3/s or mol/m^2/s)
R	universal gas constant (=8.314 J/mol/K)
r_c	rate of acetate consumption (mol/s)
$r_{CO,ad/ox}$	adsorption / oxidation rate of CO (1/s)
Re	Reynolds number (= vd/ν)
R_{el}	electrode radius (m)
R_{in}	cell radius at the inlet of the cyclone cell (m)
$r_{OH,ad/de}$	adsorption / desorption rate of OH (1/s)
r_p	rate of acetate production (mol/s)
R_u	uncompensated electrolyte resistance (Ω)

S	parameter sensitivity matrix
Sc	Schmidt number ($= \nu/D$)
Sh	Sherwood number ($= kR_{el}/D$)
T	temperature (K)
t	time
t_-	anion transference number
V	reactor volume (m^3)
$\dot{V}$	volume flow rate (m^3/s)
v_ϕ	tangential velocity at the electrode radius (m/s)
W	weight factor in the objective function
X	biomass (kg)
Z	electrochemical impedance (Ω)
z	coordinate perpendicular to the electrode surface (m)
z	number of transferred electrons

Chapter 1

Introduction[1]

1.1 Motivation

The word 'kinetics' originates from the ancient Greek word κινησις(kinesis) which means movement. In the fields of chemistry and chemical engineering the term kinetics refers to describing the rate of chemical reactions, i.e. the time it takes for one substance to convert chemically into another substance. Within chemical kinetics, microkinetics and macrokinetics are distinguished. Microkinetics are only concerned with the conversion rates of chemical species whereas macrokinetics additionally aim at describing the effects of heat and mass transport on these conversion rates. In this thesis, kinetics will be interpreted broadly as macrokinetics because the vast majority of electrochemical processes of practical relevance are limited by mass transfer to some degree.

The hallmark of electrochemistry is the direct link between reaction rate and electrical current that is described by Faraday's law:

$$I = r z F \tag{1.1}$$

The current I that results from an electrochemical reaction equals the product of the reaction rate r, the number of transferred electrons z, and the Faraday constant F. From this simple relation it is evident that the kinetics are key factors for the performance of any electrochemical process or device. Therefore, understanding electrochemical reaction kinetics is of general scientific interest and has high practical relevance for devices such as fuel cells, electrolysers, and

---

[1]Parts of this chapter have been published in F. Kubannek, C. Moß, K. Huber, J. Overmann, U. Schröder, U. Krewer, Concentration Pulse Method for the Investigation of Transformation Pathways in a Glycerol-Fed Bioelectrochemical System, Frontiers in Energy Research 6 (2018) and in F. Kubannek, U. Krewer, Modelling and parameter identification for a biofilm in a microbial fuel cell, submitted to Chemie Ingenieur Technik (2018).

batteries. These technologies are often seen as key components for the transition to an economy that is powered by renewable energy sources because they allow to store large quantities of electrical energy chemically.

Knowledge-based design and systematic improvement of electrochemical processes will only become possible in future if the underlying reaction kinetics are well-understood. At the same time, a better understanding of reaction kinetics paves the way towards a deeper knowledge of electrochemical reactions and can help to elucidate fundamental questions.

A broad range of methodologies and approaches are available to elucidate electrochemical reaction kinetics. Dynamic methods are especially valuable for understanding complex electrochemical systems and for identifying their kinetics because the corresponding response signals contain significant information on the state of the electrode and allow for the decoupling of phenomena with different time constants. They have been successfully applied to various electrochemical systems. Dynamic methods with sinusoidal change in current, e. g. electrochemical impedance spectroscopy or nonlinear frequency response analysis have been used to identify the kinetics e g. of methanol electrooxidation [1] and oxygen electroreduction [2]. Step-wise changes in current or concentration [3] have been used to study electrochemical kinetics and interaction of reaction and transport [4].

The combination of electrochemical methods and measurement of bulk concentration and surface species led to the development of various *in operando* techniques that have been shown to yield additional insights into electrode processes [5, 6]. However, usually not all state variables and quantities of interest in an electrochemical system can be accessed experimentally even with advanced analytical techniques. Modelling and simulations are suitable tools to investigate mechanisms, extract kinetic parameters and observe state variables that cannot be directly measured in electrochemical systems.

Dynamic electrochemical methods are frequently coupled with *in operando* analytical techniques and also with model-based analysis of electrochemical systems. In contrast, the combination of all three approaches is not very common. Thus this thesis is based on the hypothesis that a combination of dynamic electrochemical methods, *in operando* techniques for the detection of chemical species, and simulations, is a feasible and advantageous approach to the determination of electrochemical reaction kinetics.

1.2 Purpose and scope of this thesis

The aim of this thesis is the development of new dynamic methodologies and extension of existing methodologies to study electrochemical reaction kinetics to probe the hypothesis formulated above. In the following paragraphs, details of the chosen approach and the structure of the thesis will be laid out.

While the overall topic of this thesis is a methodological one, the developed strategies will be applied to four specific example systems which are the carbon-monoxide electrooxidation, methanol electrooxidation, acetate electrooxidation, and glycerol electrooxidation. The first two reactions are catalysed by conventional platinum ruthenium catalysts, the latter two by microorganisms in bioelectrochemical systems. For each of these electrochemical processes it will be demonstrated how the combination of dynamic current, voltage and concentration measurements yields additional insights. It will be demonstrated that concentration measurements can be used to reliably parametrise physical models of electrochemical reaction kinetics and to elucidate the effects of transport phenomena.

The reactions investigated in this thesis are discussed in ascending order of complexity of the underlying reaction mechanisms, starting from carbon-monoxide and methanol, that contain one carbon atom per molecule, via acetate, that contains two carbon atoms, ending at glycerol that contains three carbon atoms. Since the mechanism of carbon-monoxide electrooxidation is comparatively well understood whereas that of bioelectrochemical glycerol is largely unexplored, the challenges and the level of detail differ between the systems.

For all example systems, the identification of parameters from experimental data consisting of current, potential, and concentration measurements is demonstrated. In table 1.1, an overview on the methods that are used in this thesis is given. It is shown how identification procedures can account for a different number of data points or different measurement accuracies using straightforward and practical approaches as well as rigorous determination of confidence intervals and parameter interactions based on likelihood methods.

In the first part of this thesis, electrochemical oxidation reactions on porous technical electrodes are addressed where mass transfer and spatial gradients within the electrodes pose a challenge. With flux-based frequency response analysis, a novel dynamic electrochemical measurement technique is introduced. Also, for the first time, Differential Electrochemical Mass Spectroscopy (DEMS) measurements

Table 1.1: Overview on the dynamic techniques and concentration measurements employed within this thesis

System	experiment	concentration measurement	simulation
CO oxidation	potential step (CA)	DEMS	yes
methanol oxidation	CV, EIS, species flux-based FRA	DEMS	yes
acetate oxidation	CV, potential step	DEMS	no
glycerol oxidation	CV, concentration pulse (CA)	HPLC	partly

CA = Chronoamperometry, DEMS = Differential Electrochemical Mass Spectrometry,
CV = Cyclic Voltammetry, EIS = Electrochemical Impedance Spectroscopy,
FRA = Frequency Response Analysis, HPLC = High Performance Liquid Chromatography.

are quantitatively linked to physical modelling of electrochemical reaction kinetics and identification of rate constants.

In the second part of this thesis, the methodologies are extended to bioelectrochemical systems. One key methodological difference to the first part arises from the fact that the reaction mechanisms are less well-known and kinetic analysis does not only require to identify the values of kinetic constants for established mechanisms but also to gain a basic understanding of the system dynamics, time constants, and pathways. To this end, the first DEMS measurements on bioelectrochemical systems are performed and analysed. Additionally, it will be shown how quantitative data can be obtained and relevant conclusions can be drawn from suitable dynamic experiments even for these complex biological systems in which mechanisms are still unclear.

In **chapter 2**, the fundamentals of electrochemical reaction kinetics will be explained. First, the Butler-Volmer equation and the most common adsorption models are derived and discussed. Subsequently, a brief introduction into electrochemical energy conversion systems is given that motivates the choice of the example reactions in part one of the thesis. In the second part of the chapter, an introduction into bioelectrochemical systems is given and kinetic models for these processes are presented and compared. Finally, analytical techniques that can be used for *in operando* studies on electrochemical reaction kinetics will be briefly reviewed to explain the choice of concentration measurement methods used within this thesis.

In **chapter 3**, CO electrooxidation in a porous Pt/Ru electrode is analysed. Potential step experiments are carried out in a DEMS cell that is constructed and characterised in detail. The DEMS allows to monitor not only the transients of current density but also the CO_2 production over time. Here DEMS measurements are quantitatively correlated to simulation results from a physical model for the CO oxidation for the first time in order to identify rate constants.

In **chapter 4**, a dynamic model for the methanol electrooxidation is verified using dynamic DEMS measurements, and kinetic parameters for the model are identified. Additionally, DEMS data is evaluated in the frequency domain for the first time and the novel method of flux-based frequency response analysis is introduced. Furthermore, a detailed analysis of the contributions of different processes to the frequency response spectra and of the capabilities and limitations of the new technique is presented.

In **chapter 5**, the first DEMS measurements of an acetate oxidising biofilm anode are presented. It will be shown that Mass Spectrometric Cyclic Voltammetry and potential step experiments allow to analyse metabolic processes in the biofilm. Using this approach, internal substrate storage processes are observed and quantified.

In **chapter 6**, glycerol electrooxidation in a bioelectrochemical system is investigated. A concentration pulse methodology is applied and responses of concentration and current are analysed. This allows to separate the conversion steps between the numerous intermediates that occur in the system. It is elucidated which intermediates are important for current production, and rate constants for the metabolism of individual intermediates are determined. Furthermore, acetate oxidation in a biofilm is analysed in detail. A physical model for biofilm growth and biofilm kinetics is developed, a systematic parameter identifiability analysis is performed, and parameters as well as their correlations are determined. It is demonstrated that dynamic experimental techniques, modelling and rigorous parameter identification can also be applied to reaction systems where the exact mechanism is not clear yet.

In **chapter 7**, main findings are summarised and conclusions are drawn.

Chapter 2

Fundamentals

In this chapter, first a brief introduction into electrochemical energy conversion systems is given that motivates the choice of the example reactions in part one of the thesis. Subsequently, basic concepts of electrochemical reaction kinetics are briefly summarised. Next, the fundamentals of bioelectrochemical systems that are subject of part two of the thesis are laid out. Finally, analytical techniques that can be used for *in operando* studies on electrochemical reaction kinetics will be briefly reviewed to explain the choice of concentration measurement methods used within this thesis. Specific literature on the electrochemical reactions that are covered in this thesis, i. e. for CO oxidation, methanol oxidation, acetate oxidation in an electroactive biofilm, and glycerol oxidation in a bioelectrochemical system, is discussed in the respective chapters.

2.1 Electrochemical systems for energy conversion

Electrochemical energy systems such as fuel cells and batteries can directly convert chemical energy into electrical energy or electrical energy into chemical energy with high efficiency. Thus they have the potential to help establishing a green, CO_2 neutral economy.

The motivations for using electrochemical energy system are a high efficiency, ease of up or down scaling, potentially low emissions and the ability for flexible operation under dynamic conditions or at partial load. The reactions that are investigated in this thesis are mainly relevant for fuel cells. Thus batteries, electrolysers, and supercapacitors will not be be discussed in detail in the following.

In figure 2.1, an overview is given on common types of fuel cells and their potential applications.

Among the various types of fuel cells, hydrogen fuel cells exhibit the highest power densities and are subject to intensive research [8]. However, high volumetric

Typical applications	Portable electronics equipment	Cars, boats, and domestic CHP	Distributed power generation, CHP, also buses
POWER in Watts	1 10 100	1k 10k 100k	1M 10M
Main advantages	Higher energy density than batteries Faster recharging	Potential for zero emissions Higher efficiency	Higher efficiency less pollution quiet
Range of application of the different types of fuel cell	DMFC	AFC	MCFC
		SOFC	
	PEMFC		
		PAFC	

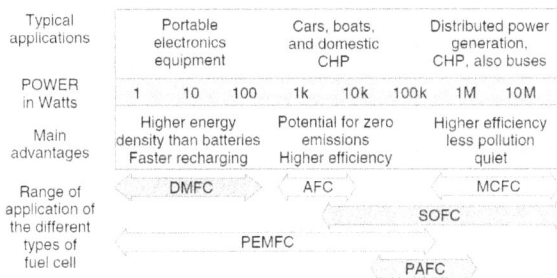

Figure 2.1: Overview on possible applications of different types of fuel cells. Reprinted from [7] with permission of John Wiley and Sons.

energy densities are not easily achieved and require storing hydrogen at conditions such as high pressure or very low temperature. This makes hydrogen storage a challenging task. Because of this, carbon containing fuels that are easier to handle are seen as promising for mobile or portable application [9]. Methanol is liquid at room temperature and can be converted in direct methanol fuel cells (DMFC) at low temperature. Larger organic molecules that contain bonds between carbon atoms can currently only be utilised at high temperatures in solid oxide fuel cells (SOFC) or molten carbon fuel cells (MCFC).

In chapter 3, the CO oxidation reaction is investigated because CO is an important intermediate in the complete oxidation of any organic molecule. Also it is known to poison the catalyst in hydrogen polymer electrolyte membrane fuel cells (PEMFC)[10, 7]. In chapter 4, the methanol oxidation reaction is investigated which is relevant for DMFC.

Bioelectrochemical systems are treated separately in the next section because many concepts and the overall state of development of the technology differ from the systems mentioned here.

2.2 Electrochemical reactions and their kinetics

In this section, basic equations that govern electrochemical reaction kinetics on solid electrodes will be explained. All these equations rely on mean-field approximations. Approaches from computational chemistry such as molecular dynamics or density functional theory are not discussed because they cannot describe processes at non-ideal technical electrodes and at realistic time and length scales yet. For the sake of simplicity, the applicability of dilute solution theory will be assumed and concentrations will be used instead of activities. First, the Butler-Volmer equation that describes the rate of electron transfer at a solid-electrolyte interface in relation to electrode potential and reactant concentrations will be derived. Next, rate equations to describe electrochemical reactions which involve adsorbed intermediates are introduced, and the Langmuir, Temkin, and Frumkin adsorption models are explained.

2.2.1 The Butler-Volmer equation

The Butler-Volmer equation is the most common way to describe electrode kinetics. In this section the equation will be derived. The section is based on the seminal book by Bard and Faulkner [11].

The net rate r of a chemical homogeneous-phase reaction, such as reaction 2.1 from species A to species B, can be described as the difference of the rate of the forward reaction r_f and the backward reaction r_b.

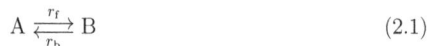

$$A \underset{r_b}{\overset{r_f}{\rightleftharpoons}} B \tag{2.1}$$

The rates of forward and backward reaction are described by formal kinetics as the product of a rate constant k and the concentration c of the reactant. For reaction 2.1 this yields the following equations:

$$r_f = k_f c_A \tag{2.2}$$
$$r_b = k_b c_B \tag{2.3}$$
$$r = r_f - r_b \tag{2.4}$$
$$= k_f c_A - k_b c_B \tag{2.5}$$

According to the Arrhenius equation, the reaction rate constants can be expressed as a function of the frequency factor k_0, the temperature T, the universal gas constant R, and the activation energy ΔG^A of the reaction:

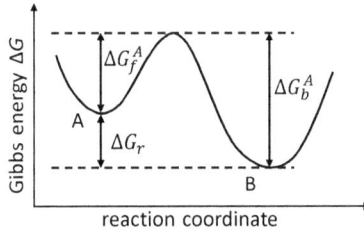

Figure 2.2: Schematic representation of Gibbs energy during a chemical reaction.

$$k = k_0 \exp\left(\frac{-\Delta G^{\mathrm{A}}}{RT}\right) \tag{2.6}$$

The concept of the activation energy is shown schematically in figure 2.2. The Gibbs energy over a theoretical reaction coordinate that represents the degree of completion of the reaction from A to B must exhibit at least two local energy minima at A and B because product and educt are stable. Between the two stable species, a transition state at a higher energy level exists. The activation energy for the forward (backward) reaction $\Delta G_{\mathrm{f}}^{\mathrm{A}}$ ($\Delta G_{\mathrm{b}}^{\mathrm{A}}$) is the difference between the Gibbs energy of the educt (product) and this transition state. Within the Transition State Theory, methods to determine ΔG^{A} and k_0 have been developed [11]. Since these approaches require many assumptions such as reversibility of reactions, which are often not met in real systems, and a good understanding of the transition state, the rate constant will be identified from experimental data in this thesis.

For heterogeneous electrochemical reactions at solid electrode surfaces such as reaction 2.7, the reaction rates depend on the surface concentrations of the species instead of the bulk concentrations:

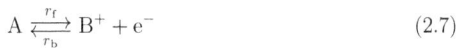

$$A \underset{r_{\mathrm{b}}}{\overset{r_{\mathrm{f}}}{\rightleftarrows}} B^+ + e^- \tag{2.7}$$

$$r = r_{\mathrm{f}} - r_{\mathrm{b}} \tag{2.8}$$

$$= k_{\mathrm{f}} c_{\mathrm{A}}^0 - k_{\mathrm{b}} c_{\mathrm{B}}^0 \tag{2.9}$$

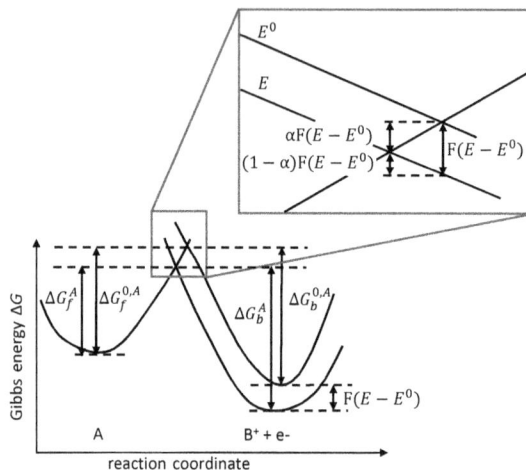

Figure 2.3: Schematic representation of Gibbs energy during a one-step, one-electron transfer electrochemical reaction at two different electrode potentials.

c_B^0 and c_A^0 denote the concentrations of A and B directly at the electrode surface. Since an electron participates in the reaction, the overall reaction Gibbs energy as well as the activation energy are functions of the potential for an electrochemical reaction. If the electrode potential is increased from the equilibrium potential E^0 to a more positive value E, the Gibbs energy of the electron on the product side is lowered. This leads to a shift in the Gibbs energy over the reaction coordinate by $F(E - E^0)$. In figure 2.3, the Gibbs energy over the reaction coordinate is compared for the two potentials. Not only the Gibbs energy of the products is reduced but also the energy of the transition state that determines the activation energies of the forward and backward reaction. The magnified section on the upper right of the figure shows how the two activation energies are affected. The activation energy of the backward reaction is increased by $(1 - \alpha)F(E - E^0)$, that of the forward reaction is decreased by $\alpha F(E - E^0)$. The factor α is called charge transfer coefficient. When the reduction reaction is defined as forward reaction, the potential change has the opposite effect, and the signs in the following equations have to be reversed. Using the Arrhenius equation, the following relation between the rate constants and the potential is established:

$$\Delta G_f^A = \Delta G_f^{0,A} - \alpha F(E - E^0) \tag{2.10}$$

$$\Delta G_b^A = \Delta G_b^{0,A} + (1 - \alpha)F(E - E^0) \tag{2.11}$$

$$k_f = k_{0,f} \exp\left(\frac{-\Delta G_f^A}{RT}\right) \tag{2.12}$$

$$k_b = k_{0,b} \exp\left(\frac{-\Delta G_b^A}{RT}\right) \tag{2.13}$$

$$k_f = k_{0,f} \cdot \exp\left(\frac{-\Delta G_f^{0,A}}{RT}\right) \cdot \exp\left(\frac{\alpha F(E - E^0)}{RT}\right) \tag{2.14}$$

$$k_b = k_{0,b} \cdot \exp\left(\frac{-\Delta G_b^{0,A}}{RT}\right) \cdot \exp\left(-\frac{(1 - \alpha)F(E - E^0)}{RT}\right) \tag{2.15}$$

Inserting these values into equation 2.9 results in the Butler-Volmer equation:

$$
\begin{aligned}
r = & c_A^0 k_{0,f} \cdot \exp\left(\frac{-\Delta G_f^{0,A}}{RT}\right) \cdot \exp\left(\frac{\alpha F(E - E^0)}{RT}\right) \\
& - c_B^0 k_{0,b} \cdot \exp\left(\frac{-\Delta G_b^{0,A}}{RT}\right) \cdot \exp\left(\frac{-(1 - \alpha)F(E - E^0)}{RT}\right)
\end{aligned}
\tag{2.16}
$$

The reaction rate can also be expressed as a current density via the Faraday equation. Another form of the equation is obtained when the potential-independent terms are combined into rate constants k_f^* and k_b^*:

$$r = c_A^0 k_f^* \cdot \exp\left(\frac{\alpha F(E - E^0)}{RT}\right) - c_B^0 k_b^* \cdot \exp\left(\frac{-(1 - \alpha)F(E - E^0)}{RT}\right) \tag{2.17}$$

At first glance, it seems that the parameters of the Butler-Volmer equation can be easily extracted from a set of experimental data containing current-potential curves at different concentrations and temperatures. However, a number of practical difficulties makes the determination of rate constants a true challenge.

The surface concentrations are not identical to their bulk values if current is flowing. For many electrochemical reactions the kinetically controlled potential region is very small [11]. How to cope with mass transfer effects will be discussed in detail for the CO oxidation reaction and the methanol oxidation reaction in chapters 3 and 4. Many electrochemical reactions are not ideal one-step, one-electron transfer reactions. Instead adsorption, desorption and the formation of intermediates need to be considered and reaction rates may not only depend on the concentrations of product and educt but also on adsorbed species on the electrode surface.

This is also the case for the reactions analysed in chapters 3 and 4. Hence, in the next section adsorption and desorption will be addressed.

2.2.2 Adsorption models

Often electrochemical reactions do not occur in one single step but in a number of elementary steps that include adsorption of the educt, reaction and desorption of the product as in reactions 2.18 to 2.20. In this case the surface concentrations in equation 2.9 have to be replaced by the relative surface concentrations θ_A and θ_B of the adsorbates, which represent the fraction of active catalyst sites covered with a and b respectively.

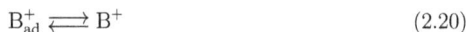

$$A \rightleftharpoons A_{ad} \tag{2.18}$$

$$A_{ad} \underset{r_b}{\overset{r_f}{\rightleftharpoons}} B_{ad}^+ + e^- \tag{2.19}$$

$$B_{ad}^+ \rightleftharpoons B^+ \tag{2.20}$$

$$r = k_f \theta_A - k_b \theta_B \tag{2.21}$$

This illustrates the need to treat not only the electron transfer itself but also the adsorption and desorption of educts, products and intermediates as an integral part of electrochemical reaction kinetics.

In the following, three common adsorption models that are often used to describe electrode processes will be explained: the Langmuir, the Frumkin and the Temkin model.

The **Langmuir adsorption model** was derived by Irving Langmuir in 1918 and is based on four main assumptions [12].

1. the adsorption process is reversible,

2. the number of available adsorption sites is limited,

3. the adsorption energy is identical on all adsorption sites,

4. the adsorbed species do not interact in any way.

With the definition of θ as the relative surface coverage with respect to the maximum number of available adsorption sites, the change of the surface coverage of a species over time can be calculated from its adsorption rate r_{ad} and desorption rate r_{des}:

$$\frac{\mathrm{d}\theta}{\mathrm{d}t} = r_{\mathrm{ad}} - r_{\mathrm{des}} \tag{2.22}$$

$$r_{\mathrm{ad}} = c_{\mathrm{A}} \cdot k_{\mathrm{ad}} \cdot (1 - \theta) \tag{2.23}$$

$$r_{\mathrm{des}} = k_{\mathrm{des}} \cdot \theta \tag{2.24}$$

In equilibrium, adsorption and desorption rate are equal:

$$r_{\mathrm{ad}}^{\mathrm{eq}} = r_{\mathrm{des}}^{\mathrm{eq}} \tag{2.25}$$

$$c^{\mathrm{eq}} \cdot k_{\mathrm{ad}} \cdot (1 - \theta^{\mathrm{eq}}) = k_{\mathrm{des}} \cdot \theta^{\mathrm{eq}} \tag{2.26}$$

$$\frac{\theta^{\mathrm{eq}}}{(1 - \theta^{\mathrm{eq}})} = c^{\mathrm{eq}} \cdot \frac{k_{\mathrm{ad}}}{k_{\mathrm{des}}} \tag{2.27}$$

Equation 2.27 is referred to as the Langmuir adsorption isotherm. In literature, the term $\frac{k_{\mathrm{ad}}}{k_{\mathrm{des}}}$ is often summarised as the adsorption coefficient B.

The **Frumkin adsorption model** allows to account for the interaction of adsorbed species, which the Langmuir adsorption model does not. Thus assumption 4 from the Langmuir model does not need to be fulfilled. The adsorption and desorption rate constants are defined as functions of the adsorption energy, which depends on the surface coverage [11]. With the assumption that relationship of rate constant and adsorption energy follows the Arrhenius equation and that the adsorption energy is linearly related to the surface coverage the following equations results:

$$k_{\text{ad}} = k_{0,\text{ad}} \cdot \exp\left(\frac{g \cdot \theta}{RT}\right) \tag{2.28}$$

$$k_{\text{des}} = k_{0,\text{des}} \cdot \exp\left(-\frac{g \cdot \theta}{RT}\right) \tag{2.29}$$

$$r_{\text{ad}} = c \cdot k_{\text{ad}} \cdot (1 - \theta) \tag{2.30}$$

$$r_{\text{des}} = k_{\text{des}} \cdot \theta \tag{2.31}$$

$$\tag{2.32}$$

$k_{0,\text{ad}}$ and $k_{0,\text{des}}$ are the reference rate constants at a surface coverage of zero. Positive values of the interaction factor g mean that attractive forces occur between the adsorbed species so that the adsorption rate increases with θ whereas the desorption rate decreases with θ. Negative values of g have the opposite effect. In equilibrium the Frumkin adsorption isotherm is described by:

$$\frac{\theta^{\text{eq}}}{(1 - \theta^{\text{eq}})} \exp\left(-\frac{2g \cdot \theta^{\text{eq}}}{RT}\right) = c^{\text{eq}} \cdot \frac{k_{\text{ad}}}{k_{\text{des}}} \tag{2.33}$$

For $g = 0$, the interactions are zero and the Frumkin isotherm is identical to the Langmuir isotherm.

The **Temkin adsorption model** was introduced in 1939 by Mikhail I. Temkin [12] and allows to account for inhomogeneous adsorption sites with different adsorption energies. Thus assumption 3 from the Langmuir model does not need to be fulfilled. Under the assumption of a uniform distribution of surface sites between an upper and lower adsorption energy, the following adsorption isotherm can be derived [11]:

$$\theta^{\text{eq}} = \frac{RT}{2g} \ln\left(\frac{k_{\text{ad}}}{k_{\text{des}}} c^{\text{eq}}\right) \tag{2.34}$$

The Temkin adsorption isotherm is valid for intermediate degrees of surface coverage, $0.8 > \theta > 0.2$. For large values of g the Frumkin isotherm is almost identical to the Temkin isotherm in this range of surface coverages. Thus it is not easy to evaluate if interactions or inhomogeneities or both are responsible for the shape of a measured adsorption isotherm [12]. This is demonstrated in a review article from Chun and Chun [13] where the determination of Frumkin, Temkin und Langmuir isotherms for hydrogen on noble metals is discussed.

The kinetic models that were deducted and discussed in this section will be used to analyse data and set up models in chapter 3 and 4. Specific literature on the model systems and reactions examined in this thesis will be summarised in the beginning of the respective chapter.

2.3 Bioelectrochemical systems

In bioelectrochemical systems (BES), microorganisms that are capable of extra-cellular electron transfer (EET) act as electrocatalysts on at least one of the electrodes. These so called electroactive bacteria (EAB) can exchange electrons between their metabolism and a solid electrode. Many EAB form biofilms directly on an electrode. The most prominent type of BES are microbial fuel cells (MFC) where an organic substrate is oxidised on the anode by EAB. MFCs and other important types of BES such as microbial electrolysis cells (MEC) will be briefly described in section 2.3.1 to provide an overview on possible fields of applications and on the conditions under which EAB should be studied. The most common methods to characterise EAB electrochemically will be discussed in section 2.3.2. A clear focus is placed on anodic biofilms because they are the main subject of the research work carried out on BES within this thesis. Finally, approaches to describe the kinetics that govern the current density and reaction rates in such biofilms will be reviewed in section 2.3.3.

Since acetate is the most common and well-studied model substrate in BES, specific examples in the following sections will cover BES that oxidise acetate. For these cells, the overall anodic half cell reaction can be formulated as:

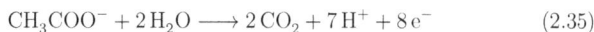

$$CH_3COO^- + 2\,H_2O \longrightarrow 2\,CO_2 + 7\,H^+ + 8\,e^- \tag{2.35}$$

The most common reactions on the cathode side are the oxygen reduction reaction in an MFC and the hydrogen evolution reaction in an MEC.

$$8\,H^+ + 8\,e^- \longrightarrow 4\,H_2 \qquad\qquad (2.36)$$

$$8\,H^+ + 2\,O_2 + 8\,e^- \longrightarrow 4\,H_2O \qquad\qquad (2.37)$$

The cathode reactions are often faciliated by non-biological catalysts. Biocathodes, where oxygen is reduced by microorganisms, have been described in literature [14, 15]. The cathodic reactions will not be covered in depth in this thesis, though.

2.3.1 Reactor concepts, transport processes and electrode materials

In this section, the most common applications and reactor concepts for BES will be introduced briefly. The most important transport processes and common electrode materials will be covered as well. Unless other sources are cited, the information is taken from a recent paper by Santoro et al. [16] who thoroughly reviewed current literature on MFCs, including cell designs, materials and applications.

In figure 2.4, schematics of four basic BES configurations are shown. Figure 2.4 (a) shows an MFC in which organic substances are oxidised at the anode and oxygen is reduced at the cathode to harvest electrical energy. MFC can be used for the treatment of waste water, a process that conventionally consumes a lot of energy. While laboratory studies on MFC often use specific bacteria, such as *Geobacter sulfurreducens* or *Geobacter anodireducens*, mixed microbial communities have to be used in practical applications. The substrate concentration as well as the ionic strength of the medium is generally low.

Figure 2.4 (b) shows an MEC in which organic substances are oxidised on the anode and hydrogen is produced on the cathode. In order to provide the necessary potential for the hydrogen evolution reaction (HER), external electric energy needs to be supplied. However, the total amount of energy is smaller than in conventional electrolysis cells. In a recent review, Kadier et al. [17] discuss various types of MEC designs including combinations of MEC and MFC or MEC and solar power.

Microbial desalination cells are an integrated way to remove salts as well as organic components from liquid streams. In these cells, a third compartment is set up between anode and cathode, separated by ion-exchange membranes. The

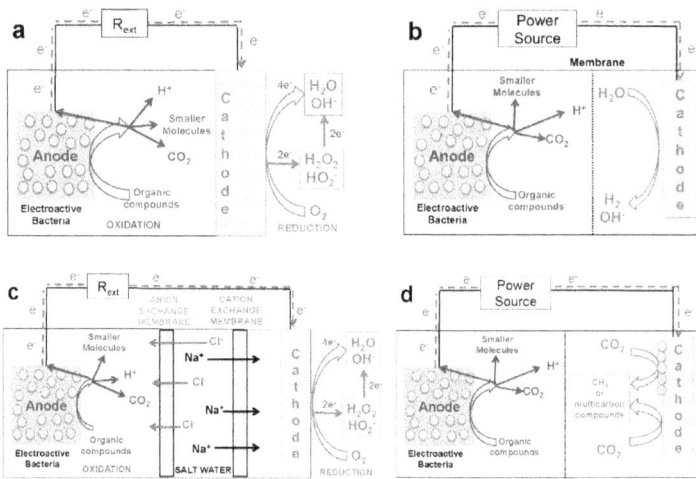

Figure 2.4: Schematic of a microbial fuel cell (a), microbial electrolysis cell (b), microbial desalination cell (c) and general microbial electrosynthesis cell (d). Reprinted from [16] (CC By 4.0).

potential difference between anode and cathode is used to generate a migrative flux of ions out of the middle compartment. Further details on this emerging application of BES are discussed in a review by Sevda et al. [18].

Another comparatively new field of application are microbial electrosynthesis cells, which are shown in figure 2.4 (d). Here, EAB on the cathode integrate electrons into their metabolism to reduce organic substances and produce value-added products.

Most BES use carbon based anode materials because of their chemical inertness and the availability of various three-dimensional structures and surface morphologies. Among the various types of materials are carbon cloth, carbon brush, carbon mesh, carbon paper, graphite plates, graphite rods, and carbonated cardboard. It has been shown that metals such as copper, gold and stainless steel can also be used if the potential is low enough to prevent corrosion [19]. In the same work it has been suggested that copper or steel can be more cost efficient than graphite because the higher conductivity allows to build large electrodes with less material.

On the cathode side of MFCs, platinum-based materials as well as platinum-group-metal-free catalysts are widely employed. It is noteworthy that the mechanisms of the HER and oxygen reduction reaction (ORR) in neutral media are less well-understood than in acidic or alkaline media even for common platinum-based materials.

While conventional fuel cells usually contain an ion-conductive membrane that separates anode and cathode, such membranes are not commonly used in BES. The reasons for this are the need for low-cost designs and low ionic resistances as well as problems resulting from membrane fouling. Instead often mechanical separators such as nylon fibers, glass fibers and ceramics are used that prevent electrical short circuits and reduce diffusion rates of oxygen from the cathode to the anode and of substrate from the anode to the cathode.

In general, transport phenomena play an important role in all BES. Popat and Torres [20] summarised the effects of the most important transport phenomena. If no other source is cited, the following points are based on their work.

- Ionic transport in BES is often slow and leads to high Ohmic losses because the ionic conductivity of the buffered electrolyte is often lower by two orders of magnitude than in Nafion or in conventional electrochemical cells where an access of highly mobile H^+ or OH^- ions is present. Furthermore, the distance between anode and cathode often increases when electrodes with a three-dimensional structure and large surface area such as carbon brushes are used. The trade-off between electrode surface area and distance between anode and cathode is discussed in [21] and [22]. Ionic transport resistances can be reduced by using buffers of high ionic strength, by abandoning membranes and separators, and by smaller distances between the electrodes.

- Crossover of species between the electrodes often reduces performance. Since many BES do not employ ion-selective membranes, oxygen (in an MFC) or hydrogen (in an MEC) can diffuse from the cathode to the anode. Oxygen impedes the activity of the EAB which are mostly anaerobic bacteria. While hydrogen can be utilised by many EAB, the crossover directly reduces the yield of the MEC. Crossover effects can be reduced by using membranes, larger distances between the electrodes or higher substrate concentrations.

- Transport of protons and the resulting pH gradients can cause large overpotentials. Migrative transport that plays a large role in conventional fuel cells is small in BES because protons represent only a small fraction of the ionic species. In a BES with a membrane the pH value on the cathode increased up to 12 [23], resulting in an additional overpotential of approximately 300 mV. In anodic biofilms the pH value is reduced by the reaction and can finally inhibit the EAB's activity. The maximum current density that is observed in matured biofilms at high substrate concentrations is most likely limited by the pH shift. Proton transport limitations can be alleviated by better mixing that leads to smaller concentration boundary layers or by increasing buffer concentrations.

- Low electric conductivity within the biofilm causes a potential gradient and reduces the available potential. However, it seems likely that transport of ionic species rather than electron conduction is limiting biofilm performance.

On the full cell level, low power densities, transport limitations, and unresolved scale up problems remain big challenges for all types of BES. For this reason the practical application of BES is still at its infancy [24, 25]. Therefore, it is claimed that a better fundamental understanding of EET is necessary to advance the field [24, 25]. In the next section, the most common electrochemical techniques to characterise electroactive biofilms and analyse their EET mechanisms are described.

2.3.2 Electrochemical characterisation

In this section, chronoamperommetry, cyclic voltammetry under turnover and non-turnover conditions on anodic electroactive biofilms will be discussed. Typical experimental results for the model substrate acetate and the model organism *G. sulfurreducens* are explained.

Anodic EAB biofilms are usually grown at a constant anode potential. The development of the current density is monitored via CA to observe biofilm formation and activity. A higher current density is usually seen as equivalent to a higher biofilm activity and performance. Figure 2.5 shows a typical CA curve during cultivation of a mixed-culture biofilm in semi-batch mode on a graphite electrode from literature [19]. After inoculation, an initial lag phase of approximately two days occurs. In this phase, the bacteria adopt to the new medium and attach to the electrode. Subsequently, a period of exponential bacterial growth on the electrode causes an exponential increase in current density. After reaching a peak seven days after inoculation, the current density decreases rapidly because the

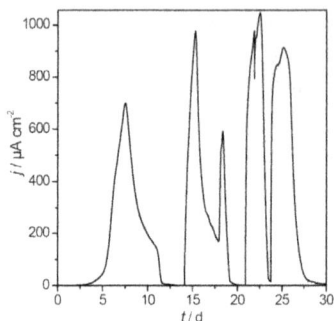

Figure 2.5: Exemplary cultivation and resulting bioelectrocatalytic current genera-tion of a secondary, acetate based electrochemically active biofilm at polycrystalline graphite in a semi-batch experiment with discontinuous addition of acetate. The biofilm was cultivated in a half-cell setup under potentiostatic control. The elec-trode potential was 0.2 V (vs. Ag/AgCl) [19]. Adapted from [19] (published by The Royal Society of Chemistry, CC BY 3.0).

substrate is depleted. The biofilm still persists in the phase of negligible current production. Therefore, the current density immediately rises again after new substrate is added on day 14. The maximum current density of approximately $1 \, \text{mA} \, \text{m}^{-2}$ is typical for mixed culture biofilms grown with acetate on non-three dimensional electrodes [20].

Two basic types of CVs are distinguished in BES: turnover and non-turnover CV [26]. A turnover CV is recorded when the biofilm is actively metabolising substrate and generating current. It is used to study the substrate oxidation characteristic of the biofilm. A non-turnover CV is recorded under substrate depletion. It is used to study the redox-properties of the biofilm itself.

In figure 2.6, exemplary CVs of *G. sulfurreducens* biofilms from literature are depicted [27]. The turnover CV curve in figure 2.6 (A) follows an s-shape. Between approximately $-0.5 \, \text{V}$ and $-0.35 \, \text{V}$ vs. Ag/AgCl, the slope of the current density increases strongly. Above $-0.35 \, \text{V}$ the increase becomes smaller and above $-0.2 \, \text{V}$ the current density reaches a plateau, indicating that the biofilm activity is not limited by the electrode potential in this range. From the maxima and minima in the slope of the turnover CV (figure 2.6 (B)), two formal potentials were determined that might correspond to the formal potentials of one or more redox proteins involved in EET. The non-turnover CVs in figure 2.6 (C) and (D) exhibit

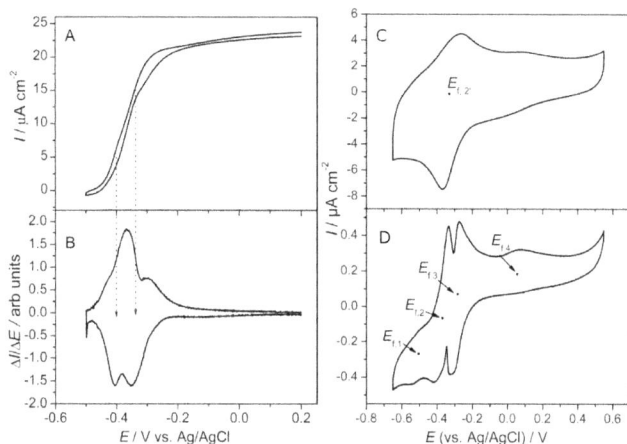

Figure 2.6: (A) Turnover CV of a *G. sulfurreducens* biofilm recorded at a scan rate of $5\,\mathrm{mV\,s^{-1}}$. (B) First derivative of the current over the potential. (C) Non-turnover CV of the biofilm recorded at a scan rate of $50\,\mathrm{mV\,s^{-1}}$. (D) Non-turnover CV at a scan rate of $1\,\mathrm{mV\,s^{-1}}$ [26]. Adopted with permission from The Royal Society of Chemistry.

a mean current density of zero, since no substrate is available and only reversible redox-factors are oxidised and reduced. At a high scan rate only one mixed formal potential can be identified because of the large capacitive currents. At a low scan rate, four formal potentials can be determined from the peak positions that are assigned to redox-proteins in the outer cell membrane, such as OmcB, OmcE, and OmcS [26]. Only the middle two formal potentials correspond to those found in the turnover CV and it can be concluded that the remaining two redox-centers do not contribute to current generation under turnover conditions.

One special challenge in the electrochemical characterization of electroactive biofilms arises from the low conductivity of the medium that is usually lower than $5\,\mathrm{mS\,cm^{-1}}$ in waste water and between 5 and $20\,\mathrm{mS\,cm^{-1}}$ for synthetic buffers. This causes a large ohmic drop and can significantly influence the voltage reading [28]. Additionally, the comparatively large capacitive currents can contribute a significant share to the total current density even at potential sweep rates as low as $1\,\mathrm{mV\,s^{-1}}$ [28].

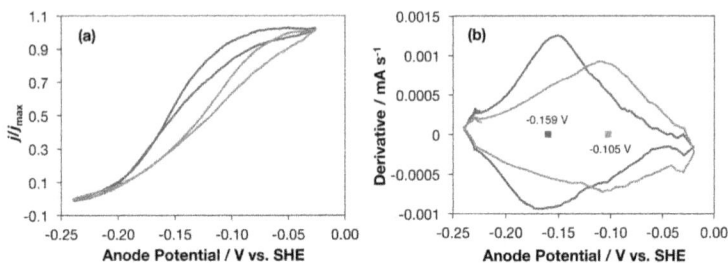

Figure 2.7: (a) Comparison of normalised CVs obtained for a fully developed *G. sulfurreducens* biofilm grown at an anode potential of -0.145 V (blue) and the same biofilm 1 h later after it has been adapted to an anode potential of -0.02 V (red). (b) Derivatives of the CVs in (a). [29]. Reprinted with permission from Wiley.

Furthermore, EAB like all other biological systems adopt to their environment and can change their electron transfer pathways even within the time frame of an electrochemical experiment. Yoho, Popat and Torres [29] demonstrated this effect by recording a turnover CV of a *G. sulfurreducens* biofilm that was grown at an anode potential of -0.145 V vs. standard hydrogen electrode (SHE), changing the anode potential to -0.02 V and recording another turnover CV after one hour of adaptation time. In figure 2.7 the recorded CVs are shown. It can be seen that a clear change in the shape of the CV and the formal potentials determined from the first derivative of the CV took place. The authors attributed this to the existence of at least two EET pathways that are activated depending on the available potential difference in order to maximise the bacteria's energy gain. The shift took place within one hour. In comparison, the time it takes to record two cycles of a CV with a typical scan rate of $1\,\mathrm{mV\,s^{-1}}$ in a 600 mV wide potential window is 40 minutes, and an EIS measurement with a minimum frequency of $1\,\mathrm{mHz}$ takes up to 20 minutes depending on the number of data points per decade. Thus it is possible that during many electrochemical characterisation experiments biofilm adaptation takes place to some degree.

Up to now, the characterisation techniques introduced in this section have helped to elucidate many characteristic features of EAB. However, in recent literature the view has been expressed that the combination of electrochemical techniques and other experimental techniques, especially online and *in-situ* methods such as microelectrodes or spectroelectrochemical methods, is necessary in order to gain detailed insights into the processes and the mechanisms that drive EET

[24]. In chapter 5, EAB are studied online with DEMS and the advantages of a multi-technique approach are demonstrated.

2.3.3 Biofilm kinetics

In this section, approaches to describe the current-potential relationship in anodic EAB biofilms will be introduced. These descriptions of biofilm kinetics will be used in chapter 6 and 6.5 to interpret experimental data and set up a model of an anodic EAB biofilm.

The most common expression to describe microkinetics in BES is the Nernst-Monod equation that was developed by Marcus et al. [30]. It is derived from the multiplicative Monod equation that describes the substrate consumption rate r of microorganisms in a liquid phase where a soluble electron donor and a soluble electron acceptor participate:

$$r = q_{max} X \frac{c_d}{c_d + K_{S,d}} \frac{c_a}{c_a + K_{S,a}} \qquad , \qquad (2.38)$$

with the concentration of the electron donor c_d, the respective half-saturation rate constant $K_{S,d}$, the concentration of the electron acceptor c_a, the respective half-saturation rate constant $K_{S,a}$, the active biomass X and the specific maximum substrate turnover rate q_{max}. In BES, the electron acceptor is not soluble but a solid electrode. To relate the potential of a solid electrode E to the concentration of a soluble electron acceptor from equation 2.38, the Nernst equation is used:

$$E = E^0 - \frac{RT}{zF} \ln \left(\frac{c_{0,a}}{c_a} \right) \qquad (2.39)$$

$c_{0,a}$ is the standard concentration of electron acceptor and E^0 is the standard potential of the electron acceptor. The concentration c_a is a hypothetical concentration value without a physical meaning and just used to derive the Nernst-Monod equation that has an empirical character like the Monod equation. When $c_a = K_{S,a}$, the rate is half of the maximum rate. The corresponding potential is defined as half-saturation potential E_{KA}, fixing the number of transferred electrons z to one implying that electrons are transferred to the electrode one-by-one.

$$E_{KA} = E^0 - \frac{RT}{F} \ln \left(\frac{c_{0,a}}{K_{S,a}} \right) \qquad (2.40)$$

Equations 2.39 and 2.40 can be rearranged for c_a and $K_{S,a}$ and substituted into equation 2.38 to obtain:

$$r = q_{max} X \frac{c_d}{c_d + K_{S,d}} \frac{\exp\left(\frac{F}{RT}(E - E^0)\right)}{\exp\left(\frac{F}{RT}(E_{KA} - E^0)\right) + \exp\left(\frac{F}{RT}(E - E^0)\right)} \tag{2.41}$$

$$= q_{max} X \frac{c_d}{c_d + K_{S,d}} \frac{1}{1 + \exp\left(-\frac{F}{RT}(E - E_{KA})\right)} \tag{2.42}$$

From the substrate consumption rate r, the current I can be determined if the share of the substrate that is used for microbial growth is known. Since this factor, the amount of biomass, and the local substrate concentration inside the biofilm are difficult to determine, the following form of the Nernst-Monod equation is often used in practice:

$$I = I_{max} \frac{1}{1 + \exp\left(-\frac{F}{RT}(E - E_{KA})\right)} \tag{2.43}$$

Here the value of I_{max} is experimentally determined and contains the effects of the remaining factors from the original equation.

Marcus et al. also developed models that describe macrokinetics in an electroactive biofilm including diffusive substrate transport and a potential distribution perpendicular to the electrode [30], migrative transport [31], and the effect of pH gradients within the biofilm [32]. However, these models are complex and it is not easy to identify parameters and to validate the model predictions quantitatively.

Rimboud et al. [33] suggested that the current-voltage characteristic of an electroactive biofilm with various redox-centers with different formal potentials could be described better by adding up the contributions of the three redox-centers to the total current:

$$I = \sum_j^N I_j \tag{2.44}$$

$$I_j = I_{max,j} \frac{1}{1 + \exp\left(-\frac{F}{RT}(E - E_{KA,j})\right)} \tag{2.45}$$

An alternative to the Nernst-Monod equation is the Butler-Volmer-Monod equation that was developed by Hamelers et al. [34]:

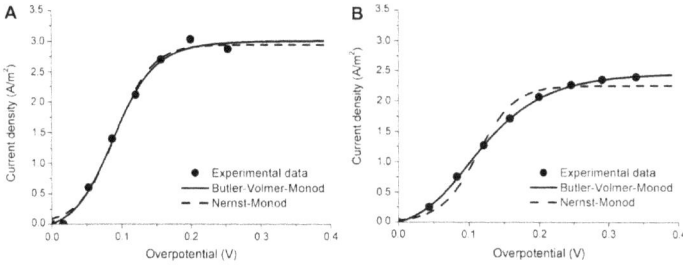

Figure 2.8: Comparison of the fit for the Nernst-Monod and the Butler-Volmer-Monod equation to two different experimental data sets [34]. Reprinted with permission from Elsevier.

$$I = I_{\max} \frac{1 - \exp(-\frac{F}{RT}\eta)}{K_1 \exp(-(1-\alpha)\frac{F}{RT}\eta) + K_2 \exp(-\frac{F}{RT}\eta) + (K_M c_d^{-1} + 1)} \tag{2.46}$$

The overpotential η is defined as $\eta = E - E^0$. $I_{\max}$ is the maximum current, K_1 and K_2 are empirical parameters. K_M is the substrate affinity constant or Michaelis-Menten constant. It is defined as the concentration where half of the maximum turnover is reached. It has been shown that this equation can describe the observable current-potential relationship better than the Nernst-Monod equation in some cases. In figure 2.8, the fit of both equations to two different sets of experimental data is compared. It can be seen that the first data set is reproduced equally well by both equations, the second data set is reproduced more accurately by the Butler-Volmer-Monod equation.

A further advantage of the Butler-Volmer-Monod equation is the fact that no net reaction rate is predicted at equilibrium potential. This is not the case in the Nernst-Monod equation that is derived from the Monod equation and can thus only describe irreversible reactions. However, the Butler-Volmer-Monod equation contains one additional empirical parameter that needs to be determined and is less commonly used [35]. Therefore, the Nernst-Monod equation will be used in chapter 6.

2.4 Electrochemical and analytical techniques for studying kinetics

In this section an overview is given on common electrochemical and analytical techniques that can be applied to investigate electrochemical and bioelectrochemical reaction kinetics. It will be briefly explained which methods and techniques are used within this thesis. A clear focus is placed on techniques that can be used for dynamic analysis and are able to provide time-resolved data.

In figure 2.9, the typical time scales of dynamic processes in electrochemical and bioelectrochemical systems and the typical time scales of common electrochemical and analytical techniques are shown. In order to investigate a specific process, its time scale needs to be covered by the electrochemical technique. An accompanying analytical technique must be capable of following the fastest changes that are induced by the process of interest.

The analytical techniques included in figure 2.9 are a selection of techniques commonly used in the field of electrochemistry and are not meant to constitute a complete overview over the vast field of analytical chemistry. The selection is mainly based on a recent review on *in operando* spectroscopy and microscopy for analysis of electrochemical processe by Choi at al. [6] and the seminal book by Bard and Faulkner [11]. The time scales or response times of analytical methods often depend on the specific cell set up, and especially in spectroscopic methods there is often a trade-off between sensitivity and measurement time so that sophisticated set ups and advanced detectors allow for higher time resolution. The response times reported in figure 2.9 are based on literature and might be significantly lower than those of commercial standard set ups.

In the following paragraphs, the techniques from figure 2.9 will be briefly described in terms of response time, ability to identify specific species, and possibility of quantification. Additionally, references containing more detailed descriptions of the techniques and their applications will be given.

- Scanning electrochemical microscopy (SECM) is a very fast technique since changes in current or potential of a very small electrode tip near the working electrode are measured directly. It allows to determine spacial profiles in the bulk phase near the working electrode. Data can be quantified, but it is not possible to directly identify individual species because only the tip

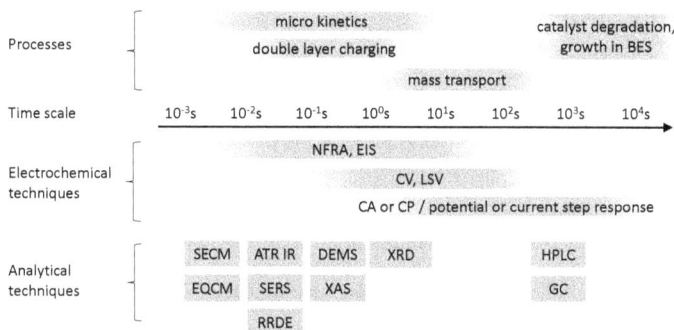

Figure 2.9: Typical time scales of processes in electrochemical and bioelectrochemical systems and of electrochemical and analytical techniques that are commonly used to study these processes.

current or potential is read. Advances in SECM were recently reviewed by Polcari and coworkers [36].

- Electrochemical quartz crystal microbalance (EQCM) can be used to monitor the changes in electrode mass resulting from the adsorption and desorption of intermediates. Like SECM it is a very fast technique and allows quantification. However, only the total change in mass of the adsorbates can be measured and different species cannot be distinguished. Further information on ECQM working principle and the application in studying electrode processes can be found in [37].

- Infrared spectroscopy (IR) is a versatile technique and widely applied to detect IR-active species. Modern spectrometers usually use Fourier transform infrared spectroscopy (FTIR). Ye et al. [38] recently published an overview on the application of *in-situ* FTIR in electrochemical systems. Here attenuated total reflection infrared spectroscopy (ATR-IR) is discussed as a promising technique for the identification of surface species. ATR-IR allows for a high time resolution but quantification is difficult. Wang and coworkers [39] give a thorough overview on the application of *in-situ* ATR-IR in electrocatalysis.

- Surface enhanced Raman spectroscopy (SERS) can detect various surface species with a high time resolution and sensitivity but limited possibilities for quantification. Many species that are not IR active, can be detected by

Raman spectroscopy. An overview on the working principle and applications of SERS can be found in [40].

- The rotating ring disk electrode (RRDE) is a technique to electrochemically detect products of reactions at a rotating disk elektrode (RDE). It does not require expensive instrumentation and allows quantification as well as fast response times [41]. Species cannot be differentiated easily and not every type of electrode can be studies in an RDE configuration.

- In differential electrochemical mass spectrometry experiments, species from a liquid electrolyte phase diffuse through a microporous membrane into a vacuum where they are detected by mass spectrometry. Thus it is limited to measuring volatile species from the bulk phase. Further information on DEMS can be found in two comprehensive reviews from Baltruschat and coworkers [42, 43]. DEMS allows quantification and direct identification of many species.

- X-Ray adsorption spectroscopy (XAS) and other related techniques such as XANES, XAFES, and XAFS can be used to measure various species in electrochemical systems *in operando* with a high spatial and temporal resolution [5]. While these approaches are very powerful and yield a lot of information, a beamline is needed as a radiation source which limits the experimental capacities in practice.

- X-Ray diffraction (XRD) can be used to follow the changes in electrode properties with a medium time resulution [44]. XRD is usually sensitive not only for the surface but for the bulk of the electrode.

- High performance liquid chromatography and gas chromatography (GC) are powerful techniques for processes with slow dynamics. They can be used to identify and quantify even trace amounts of almost any substance. GC and HPLC both require samples to be taken from a bulk phase and cannot be used to study surface phenomena.

For further insight into well established electrochemical standard techniques such as cyclic voltammetry (CV), linear sweep voltammetry (LSV), electrochemical impedance spectroscopy, chronoamperommetry (CA), chronopotentiometry (CP), and high performance liquid chromatography, the reader is referred to textbooks such as [11] and [45]. Non-linear frequency response (NFRA) is not used in this thesis and not discussed in detail here.

In this thesis, EIS, CV and potential steps will be applied to cover the wide range of time scales of the processes in the examined systems. In chapters 3 to 5, electrochemical techniques will be complemented by DEMS measurements to determine concentrations of volatile reaction products with a high time resolution. In chapter 6, the concentrations will be determined by HPLC to analyse the slower dynamics in the bioelectrochemical systems. Both methods allow to quantify the amount of detected species and provide a good basis for the simulations. Neither HPLC nor DEMS detect adsorbed species directly on the electrode surface but they detect dissolved species from the bulk phase. While this is a disadvantage for studying reaction mechanisms on ideal flat surfaces, it can be seen as an advantage for studying technical electrodes or bioelectrochemical electrodes where surfaces are not well-defined and difficult to access.

Part 1 – Electrochemical oxidation reactions

Chapter 3

CO oxidation on a Pt/Ru catalyst[1]

CO oxidation is chosen as the first example system to apply the approach that was lined out in the introduction because the reaction mechanism for CO is less complex than for other organic molecules, and physical models for the surface processes are already available. This renders CO oxidation a suitable system to establish and test the combination of *in operando* measurements, dynamic electrochemical experiments and modelling. Potential steps are applied as a dynamic technique because the system changes from one steady state to another making data analysis less complex than in case of e.g. CV where the system is always in a transient state. Since the dynamics of the reaction are expected to occur on a fast time scale and the product CO_2 is volatile, DEMS is selected as a suitable analytical *in operando* technique.

3.1 Introduction

DEMS is a powerful technique for analysing volatile products and intermediates of electrochemical reactions online and directly at the electrode. First cells were designed by Wolter and Heitbaum [46] who deposited a porous electrode directly on a porous Teflon membrane through which reaction products could enter a differentially pumped vacuum chamber and, subsequently, be detected by the mass spectrometer within approximately 0.1 seconds. Since DEMS is a quantitative technique and volatile species can be detected with an extremely short time delay, it seems highly suitable for analysing reaction kinetics and dynamic processes.

The performance of a DEMS device, however, strongly depends on the cell design which has to be adopted to the desired application. Therefore, numerous

---

[1]Parts of this chapter have been published in F. Kubannek, U. Krewer, A Cyclone Flow Cell for Quantitative Analysis of Kinetics at Porous Electrodes by Differential Electrochemical Mass Spectrometry, Electrochimica Acta 210 (2016) 862–873.

electrochemical DEMS cells have been designed for the purpose of solving specific scientific problems.

Since there is no convection in the classical DEMS cells, the mass transfer is not well-defined making it difficult to analyse concentration dependent adsorption and reaction processes that play an important role in technical electrodes. Fujihira and coworkers [47] included a rotating rod above the working electrode to increase mass transport. Wasmus and coworkers [48] used a rotating porous electrode which they placed near to the PTFE membrane. This approach led to increased convection but the flow was still not well-defined and only a small fraction of the products was transported through the solution and the membrane into the vacuum system. Another approach for generating defined convection is to use a rotating inlet system resembling a Rotating Disk Electrode [49]. In this case, a rotating vacuum feed through is required. More recently, a wall-jet configuration was suggested [50] which led to defined convection at the electrode but quantitative kinetic measurements were not possible due to flow and mass transfer limitations.

Baltruschat and coworkers developed a thin-layer cell [51, 52] for the use of massive electrodes under electrolyte flow. In such cells, the porous PTFE membrane is separated from the bulk electrode by a thin layer of electrolyte through which reaction products have to diffuse before they can be detected. Later on, dual thin layer cells [53, 54, 55, 56] were developed which contain two liquid compartments: one reaction compartment and one detection compartment. Through this separation the reaction cannot be influenced by concentration changes because of transport of reactants through the PTFE membrane into the vacuum. The drawback of the thin-layer and double-thin layer cell is, however, the increased response time of about 1-2 seconds which results from the additional transport step from the electrode through the electrolyte to the PTFE membrane. DEMS has also been combined with other analytical techniques such as Electrochemical Quartz Crystal Microbalance [53] or Fourier Transform Infrared Spectroscopy [57]. A double-band-electrode channel cell was presented by Abruña and coworkers [58] which featured a detection electrode for detecting non-volatile species. Also a number of cells featuring small pinhole or capillary inlets into the vacuum have been designed for the purpose of examining single crystal electrodes [59, 52, 60] or for scanning DEMS measurements [61].

A better understanding of processes in porous technical electrodes has a high practical and economic relevance. It is widely acknowledged that smooth model electrodes and porous electrodes differ in their electrochemical properties. Nevertheless there are only few works in which DEMS is utilised for analysing technical electrodes: Pérez-Rodríguez et al. [62] examined high surface area electrocatalyst on a gas diffusion layer to analyse the effect of a microporous diffusion layer on transport using a classical DEMS cell with no convection. The same cell was also used to examine the effect of surface modifications on metallic mesoporous catalysts [63]. Niether and coworkers examined the oxidation of different fuels at high temperatures by DEMS using gas diffusion electrodes of high temperature PEMFC[64]. Seiler et al. examined the kinetics of methanol and CO adsorption by DEMS using a fuel cell flow field as an electrode and a separate detection compartment at the outlet of the flow field [65]. However, the response times of these setups are quite high which is disadvantageous for analysing dynamic processes and especially kinetics.

Nevertheless, the fast response times that could be attained by the classical DEMS cells make DEMS an attractive tool for analysing micro- and macrokinetics of electrochemical reactions quantitatively. Early work on dynamic processes by Heitbaum and coworkers [66] has already demonstrated the feasibility of dynamic analysis by DEMS for the oxidation of formic acid on Platinum. Furthermore it has been shown before that dynamic modeling of reaction kinetics can yield new insights into the behavior of reaction systems, for example in case of the oxidation of methane in a catalytic monolith, i.e. a non-electrochemical reactor [67]. Zhang and coworkers [68] proposed a model for a DEMS thin-layer flow cell which was used for parametrizing a model for CO bulk oxidation on a smooth platinum electrode. In their publication no DEMS data was incorporated into the modelling, though.

Until now, DEMS has not been used for identifying and parametrising macro kinetic models of electrochemical reactions. In this chapter, a method for quantitatively determining rate constants for reactions on porous electrodes from dynamic DEMS measurements will be introduced.

As mentioned above DEMS cells should be tailored to the experimental study. The main requirements for the goal stated above are as follows. When analysing kinetics and formulating rate expressions of electrochemical reactions, the rate determining step can depend on the concentration of reactants. This is the case for example if an adsorption step following the Langmuir, Frumkin or Temkin adsorption is rate determining. Furthermore, there is evidence that for example the proportion of completely oxidised products of the methanol oxidation reaction on certain catalysts is influenced by desorption and readsorption of intermediates [69]. Clearly, a good understanding of the concentration distribution at the surface is important to distinguish the influence of diffusion limitations and reaction limitations. Thus a defined - preferably homogeneous - concentration boundary layer is the first requirement for the cell. For many of the existing DEMS cells the concentration profile at the electrode surface is not very reproducible or quite inhomogeneous [70, 59]. The second important requirement is a short response time of the mass spectrometer (MS) signal. It is essential for analysing fast reaction steps and applying high scan rates.

Here, first a new DEMS cyclone flow cell for analysing kinetics of technical electrodes quantitatively will be introduced. Next, the flow behaviour and especially the concentration boundary layer by computational fluid dynamics (CFD) simulations and experiments will be analysed. Furthermore calibration measurements for CO_2 are presented. Results of the CFD simulations and the calibration measurements will be utilised for setting up a physical model that quantitatively describes the oxidation of CO on a carbon supported Pt/Ru catalyst. The kinetic parameters will be identified using experimental data obtained with the DEMS cyclone flow cell.

3.2 Experimental

3.2.1 Cell design

The design of the cyclone flow DEMS cell is depicted in figure 3.1. The solution enters the cell through a 2 mm diameter hole tangential to the cell wall (7), moves downwards along the cell walls in a vortex flow circulating over the working electrode (9) at the bottom of the cell, and rises up again in the middle of the cell. A detailed analysis of the flow is presented in section 3.4.1. The fluid leaves the cell through the outlet on top (5). The counter electrode (6) is placed downstream of the working electrode to prevent reaction products from the counter electrode to reach the working electrode. The working electrode is deposited directly onto a porous PTFE membrane at the bottom of the cell through which volatile species can enter the vacuum system. This leads to a minimal response time because the transport path of species produced at the electrode is short. The membrane is supported by a stainless steel frit (1) to withstand the pressure difference between

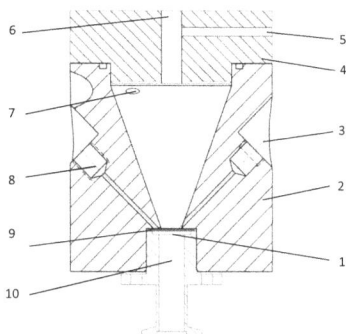

Figure 3.1: Design of the cyclone flow DEMS cell, including: (1) stainless steel frit,(2) cell body from Kel-F, (3) connection to reference electrode, (4) cell cap from Kel-F, (5) electrolyte outlet, (6) connection to counter electrode, (7) tangential electrolyte inlet, (8) Pt-wire as current collector, (9) working electrode on porous PTFE membrane and gasket, (10) connection to vacuum system / MS.

the liquid compartment and the vacuum system beneath (10). Closely above the membrane a Luggin capillary leading to the reference electrode (3) is placed. The cell body is made from PCTFE (Kel-F), seals are made from Viton. The upper cell radius is 60 mm, the angle of the cone walls is 40 degrees, and the lower cell radius is 0.5 cm resulting in an electrode area A_{el} of 0.785 cm^2.

3.2.2 Residence time measurements

As part of the flow analysis, residence time distributions were recorded: deionised water was pumped from a reservoir through the cyclone cell and the conductivity of the solution at the outlet was monitored continuously by a conductivity meter (S230, Mettler Toledo). Then a three-port-valve was switched rapidly to change to a different reservoir and hydrochloric acid solution was pumped through the cell. The experiment was stopped when the conductivity at the outlet reached that of the hydrochloric solution. Finally, the curves were normalised and corrected for the delay caused by the tubing between the reservoir and the cell.

3.2.3 Differential Electrochemical Mass Spectrometer setup

The setup consists of two vacuum chambers separated by an adjustable valve. The first chamber, which is connected to the cyclone flow cell, is evacuated by a turbo-molecular pump (Hipace300, 260 L/s, Pfeiffer Vacuum) backed by a membrane vacuum pump (MVP 015-4, Pfeiffer Vacuum). The second chamber which contains the mass spectrometer (Pfeiffer QMG220 M1 quadropole mass spectrometer with secondary ion multiplier) is evacuated by a second turbo-molecular pump (HiPace 80, 67 L/s, Pfeiffer Vacuum) backed by a rotary vane pump (Duo 5M, Pfeiffer Vacuum). The pressure in both compartments is monitored by two vacuum pressure sensors. A Gamry Reference 3000 potentiostat is used for the electrochemical experiments. A LABVIEW program specially developed for this purpose is used to record MS and potentiostat data. Ten data points are collected per second to reduce the noise in the MS signal.

3.2.4 Solutions and electrode preparation

Solutions were prepared with ultrapure water (Millipore Milli-Q 18.2 MΩcm). 0.25 M Perchloric acid (Sigma Aldrich, ACS grade) was used as a supporting electrolyte for electrochemical measurements. CO-saturated solutions were prepared by bubbling CO (99.97 %, Westfalen AG) for at least 20 minutes through the electrolyte. To produce the porous electrodes, the catalyst Pt/Ru (1:1) on 40% w/w carbon black (Johnson Matthey, HiSpec 10000) was mixed with ultrapure water, Isopropanol (VWR, HPLC grade), and Nafion solution (5% w/w of catalyst weight, Qintech NS05), ultrasonicated for 20 minutes in ice water and then spray coated with a stream of Nitrogen (99.999% Westfalen AG) directly onto the PTFE membrane (Pall Membranes, specified pore size 0.2 μm). To determine the thickness and porosity of the Pt/Ru catalyst layer, the cross section of a membrane with catalyst layer of the same type was examined by scanning electron microscopy (SEM). The thickness of the membrane is approximately $\delta^{\mathrm{M}} = 60\,\mu m$ and the thickness of the catalyst layer is approximately $\delta^{\mathrm{el}} = 25\,\mu m$. From the catalyst loading and the densities of platinum (21.5 g/cm^3) ruthenium (12.5 g/cm^3), carbon black (2 g/cm^3) and Nafion (1.8 g/cm^3), a porosity of $\epsilon^{\mathrm{el}} = 0.92$ is calculated according to equations 3.1 and 3.2.

$$\epsilon_{\mathrm{el}} = 1 - \frac{V_{\mathrm{solid}}}{V_{\mathrm{el}}} \tag{3.1}$$

$$V_{\mathrm{solid}} = \frac{m_{\mathrm{catalyst}}}{\rho_{\mathrm{catalyst}}} + \frac{m_{\mathrm{carbon}}}{\rho_{\mathrm{carbon}}} + \frac{m_{\mathrm{nafion}}}{\rho_{\mathrm{nafion}}} \tag{3.2}$$

In the literature porosities of 80% ±5 have been reported for catalyst layers in hot-pressed membrane electrode assemblies when there was no carbon in the catalyst. Thus, the porosity lies within a reasonable range. In the same way the porosity of the expanded PTFE membrane is calculated from the thickness and weight of the membrane and the density of non-porous PTFE $(2.2\,\mathrm{g/cm^3})$ as $\epsilon^M = 0.72$.

The counter electrode was made from a platinum wire. Measurements were conducted at room temperature of $25 \pm 0.5\,^\circ\mathrm{C}$. A commercial saturated Silver/ Silver-Chloride electrode (Meinsberg- Elektroden) was used as reference electrode. All potentials are, however, reported with respect to the potential of a reversible hydrogen electrode.

For the MS calibration, CO_2 (99.999 %, Westfalen AG) was leaked into the first vacuum chamber through an adjustable leak valve.

3.3 Modelling

3.3.1 Computational fluid dynamics

The CFD simulations of the cyclone flow cell are carried out using the commercial finite volume code FLUENT 15.07. A pressure-based solver was selected because of the incompressible nature of the flow. The partial differential equations were discretised using the SIMPLEC (Semi-Implicit Method for Pressure-Linked Equations-Consistent) method for pressure - velocity coupling and the Second Order Upwind scheme for pressure and momentum interpolation. The SST k-ω turbulence model was chosen because of the rather low Reynolds numbers near the bottom of the cyclone cell. It should be noted that the Reynolds numbers of the flow in gas cyclones for the separation of particles are usually higher so that different turbulence models are preferred for such flows [71]. A mesh independence study was performed and a 900,000 element unstructured mesh of tetraeders including inflation layers at the walls was used for the simulations. Boundary conditions are constant velocity at the inlet, constant pressure at the outlet and no-slip at the walls. For the flow simulation, the Luggin capillary is not taken into account.

Also, transient calculations were performed to validate the CFD results by comparing them to experimental residence time distributions. After the fluid flow was fully developed, the boundary condition at the inlet was changed and a second species was introduced. Monitoring the concentration of the second species at the outlet over time yields the residence time distribution.

When analysing reaction kinetics, the concentration boundary layer in the diffusion limited case is of primary interest. The diffusion limited conditions are modelled by introducing a species A (5% w/w, $D = 2.03 \cdot 10^{-9}\,\mathrm{m^2/s}$ based on dissolved carbon monoxide. 5% w/w exceeds the solubility of carbon monoxide at ambient pressure; such a high value is used to avoid numerical errors) at the inlet that reacts at the bottom of the cell to a species B with the same density and diffusion properties. The rate constant of the surface reaction is set very large (10^{15}) so that the concentration of species A at the surface approaches zero. This scenario allows to use Fluent's built in reaction module and is similar to a diffusion limited electrochemical reaction since the conversion rate only depends on the mass transfer towards the surface. Diffusion limited reaction rates - i.e. mass transfer rates - are calculated from the amount of species A converted to B and can easily be expressed in terms of a limiting current density.

3.3.2 Model for CO oxidation at a porous electrode

In this section a dynamic model of the porous Pt/Ru electrode described in section 3.2.4 is established. The outputs - current and the amount of CO_2 entering the DEMS vacuum system - can be directly compared to experimental data which were obtained for CO bulk oxidation. By matching experimental and simulated data, rate constants are determined.

CO oxidation on Pt/Ru follows a two step mechanism consisting of the dissociative adsorption of water and the reaction of adsorbed OH and adsorbed CO to CO_2. The first of these steps is reversible.

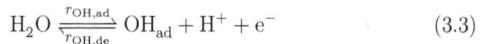

$$H_2O \underset{r_{OH,de}}{\overset{r_{OH,ad}}{\rightleftharpoons}} OH_{ad} + H^+ + e^- \tag{3.3}$$

$$CO \xrightarrow{r_{CO,ad}} CO_{ad} \tag{3.4}$$

$$CO_{ad} + OH_{ad} \xrightarrow{r_{CO,ox}} CO_2 + H^+ + e^- \tag{3.5}$$

The CO oxidation kinetics on Pt/Ru are modeled similarly to [72] where the oxidation of pre-adsorbed CO on a Pt/Ru catalyst without carbon is examined in an RDE setup. The relative surface coverages of OH and CO change over time by adsorption, desorption and oxidation:

$$\frac{d\theta_{OH}}{dt} = r_{OH,ad} - r_{OH,de} - r_{CO,ox} \tag{3.6}$$

$$\frac{d\theta_{CO}}{dt} = r_{CO,ad} - r_{CO,ox} \tag{3.7}$$

θ_{OH} and θ_{CO} denote the relative surface coverages of OH and CO, $r_{OH,ad}$ and $r_{OH,de}$ are the OH adsorption and desorption rates respectively (reaction 3.3). $r_{CO,ad}$ is the CO adsorption rate (reaction 3.4) and $r_{CO,ox}$ is the final oxidation step (reaction 3.5).

The Frumkin/Temkin adsorption isotherm is assumed for OH [72], Langmuir adsorption for CO [73]. Significant exchange rates between dissolved and adsorbed CO were found in a study by Heinen et al. [74]. This process is not taken into account here because in the chosen mean field approximation with area-averaged values of the relative coverage, the overall oxidation rate is not influenced by such an exchange step. Other experimental studies [75, 76] show that CO desorption which would yield empty adsorption sites can be neglected for experiments lasting only several minutes. The adsorption and reaction rates are thus described by equations 3.8 to 3.11.

$$r_{OH,ad} = k_{OH,ad}(\theta) \cdot (1 - \theta_{OH} - \theta_{CO}) \cdot \exp\left(\frac{\alpha_{OH} \cdot F \cdot E}{R \cdot T}\right) \tag{3.8}$$

$$r_{OH,de} = k_{OH,de}(\theta) \cdot \theta_{OH} \cdot \exp\left(-\frac{(1 - \alpha_{OH}) \cdot F \cdot E}{R \cdot T}\right) \tag{3.9}$$

$$r_{CO,ox} = k_{CO,ox}(\theta) \cdot \theta_{OH} \cdot \theta_{CO} \cdot \exp\left(\frac{\alpha_{CO} \cdot F \cdot E}{R \cdot T}\right) \tag{3.10}$$

$$r_{CO,ad} = k_{CO,ad}(\theta) \cdot (1 - \theta_{OH} - \theta_{CO}) \cdot c_{CO} \tag{3.11}$$

Water concentration is assumed not to change during the reaction and thus included into the rate constant because there is a great excess of water. For Frumkin/Temkin adsorption conditions, the rate constants depend on the relative surface coverages and the interaction / symmetry factors g and β:

$$k_{OH,ad} = k_{0,OH,ad} \cdot \exp(\beta_{OH} g_{OH} \theta_{OH}) \tag{3.12}$$

$$k_{OH,de} = k_{0,OH,de} \cdot \exp((1 - \beta_{OH}) g_{OH} \theta_{OH}) \tag{3.13}$$

$$k_{CO,ox} = k_{0,CO,ox} \cdot \exp(\beta_{CO} g_{CO} \theta_{CO} + \beta_{OH} g_{OH} \theta_{OH}) \tag{3.14}$$

It has already been pointed out that mass transfer inside porous electrodes plays an important role in a DEMS setup where the electrode is directly deposited on the PTFE membrane [42, 49]. Considering the results of the CFD simulations which will be discussed in section 3.4.1, only transport processes and gradients in the direction perpendicular to the electrode are taken into account. It is assumed that transport inside the boundary layer and the porous media is of purely diffusive nature. Thus the local concentrations of CO (c_{CO}) and CO_2 (c_{CO_2}) in the electrolyte boundary layer are described by the following partial differential equations:

$$\frac{\partial c_{CO}}{\partial t} = D_{CO,\text{electrolyte}} \frac{\partial^2 c_{CO}}{\partial z^2} \tag{3.15}$$

$$\frac{\partial c_{CO2}}{\partial t} = D_{CO2,\text{electrolyte}} \frac{\partial^2 c_{CO2}}{\partial z^2} \tag{3.16}$$

In the catalyst layer the porosity and the reaction are taken into account:

$$\frac{\partial c_{CO}}{\partial t} \cdot \epsilon^{el} = D_{CO,\text{cat}}^{\text{eff}} \frac{\partial^2 c_{CO}}{\partial z^2} - \frac{r_{CO,\text{ad}}}{A_{el} \cdot \delta^{el}} \cdot \frac{N_{\text{surface}}}{N_A} \tag{3.17}$$

$$\frac{\partial c_{CO2}}{\partial t} \cdot \epsilon^{el} = D_{CO2,\text{cat}}^{\text{eff}} \frac{\partial^2 c_{CO2}}{\partial z^2} + \frac{r_{CO,\text{ox}}}{A_{el} \cdot \delta^{el}} \cdot \frac{N_{\text{surface}}}{N_A} \tag{3.18}$$

The transport through the porous membrane is described by equations 3.19 and 3.20:

$$\frac{\partial c_{CO}}{\partial t} \cdot \epsilon^{M} = D_{CO,M}^{\text{eff}} \frac{\partial^2 c_{CO}}{\partial z^2} \tag{3.19}$$

$$\frac{\partial c_{CO2}}{\partial t} \cdot \epsilon^{M} = D_{CO2,M}^{\text{eff}} \frac{\partial^2 c_{CO2}}{\partial z^2} \tag{3.20}$$

$D_{i,j}$ is the diffusion coefficient of species i, i.e. CO or CO_2 in j $\in$ {electrolyte, electrode, membrane}; z is the coordinate perpendicular to the electrode. N_{surface} is the total number of available surface sites and A_{el} is the electrode area. As above, it is assumed that the water concentration does not change significantly through the reaction. Thus, water transport is not modelled.

The diffusion coefficients of CO and CO_2 in the electrolyte are assumed to be equal to those in water: $D_{CO,H_2O} = 2.03 \cdot 10^{-9}\,\text{m/s}^2$ at $25\,°C$ [77], $D_{CO_2,H_2O} = 1.92 \cdot 10^{-9}\,\text{m/s}^2$ at $25\,°C$ [77]. The effective diffusion coefficients in the porous electrode and membrane are assumed to follow the Bruggeman equation:

$$D_{i,j}^{\text{eff}} = D_{i,\text{H}_2\text{O}} \cdot \epsilon_j^{1.5} \tag{3.21}$$

With porosities ϵ_j being that of the electrode or membrane. The thicknesses of membrane and electrode were determined by SEM as described above, whereas the thickness of the concentration boundary layer is determined from CFD simulations.

As boundary conditions, the concentrations of all species are fixed to zero at the membrane/vacuum interface and to their bulk values at the outer edge of the concentration boundary layer:

$$c_{\text{CO}}|_{z=0} = c_{\text{CO,bulk}} \tag{3.22}$$

$$c_{\text{CO2}}|_{z=0} = c_{\text{CO2,bulk}} \tag{3.23}$$

According to the experimental conditions (compare section 3.2), the CO bulk concentration is set to saturation concentration ($1.1\,\text{mol/m}^3$ at $20\,°\text{C}$) whereas the CO_2 bulk concentration is set to zero.

At the membrane/vacuum interface all concentrations are fixed to zero. The initial values of all concentrations and surface coverages are equal to the steady state values at a fixed potential in the simulations. The steady state values were found by simulating the model for a long time starting from arbitrary initial values.

The double layer is modelled as a capacitor with constant capacitance C_{dl} according to [11]. The charge stored in the double layer is the integral of the total current minus the Faradaic current from reactions 3.3 and 3.5:

$$\frac{\mathrm{d}Q_{\text{dl}}}{\mathrm{d}t} = \frac{\mathrm{d}E}{\mathrm{d}t} \cdot C_{\text{dl}} = I - I_{\text{reaction}} \tag{3.24}$$

$$I_{\text{reaction}} = Q_{\text{m}} \cdot (r_{\text{CO,ox}} + r_{\text{OH,ad}} - r_{textOH,de}) \tag{3.25}$$

With $Q_{\text{m}} = \text{F} \cdot N_{\text{surface}}/N_{\text{A}}$ being the charge of one monolayer of adsorbed molecules on the catalyst surface.

The apparent potential between the working electrode and the reference electrode E_{external}, which can be measured in an experiment by a potentiostat, is slightly larger due to the uncompensated solution resistance R_{u}:

$$E_{\text{external}} = E + R_{\text{u}}I \tag{3.26}$$

The flow $\dot{n}$ of CO_2 and CO entering the vacuum system is calculated from the concentration gradient at the membrane/vacuum interface according to equations 3.27 and 3.28 where A_{el} denotes the electrode area.

$$\dot{n}_{MS,CO_2} = -D^{eff}_{CO_2,M} \cdot A_{el} \cdot \left(\frac{\partial c_{CO_2}}{\partial z} \right) \Big|_{mem/vac} \tag{3.27}$$

$$\dot{n}_{MS,CO} = -D^{eff}_{CO,M} \cdot A_{el} \cdot \left(\frac{\partial c_{CO}}{\partial z} \right) \Big|_{mem/vac} \tag{3.28}$$

It is assumed that transport inside the vacuum system is fast and that there is no holdup [42].

To solve this set of equations, the concentration boundary layer inside the electrolyte from section 3.4.1, the electrode, and the membrane are discretised in z-direction, perpendicular to the electrode surface, with the finite volume method assuming piecewise linear profiles. Each layer is discretised into 40 elements and the resulting ordinary differential equations are solved numerically by Matlab solver ode23t.

The values of geometric parameters used for the simulation are the same as in the experimental setup. The values and identification procedure for the kinetic parameters are discussed in the results section 3.4.3.

3.4 Results and discussion

3.4.1 Flow analysis of the cyclone cell

Sundmacher [78] proposed a cyclone flow cell for investigating gas diffusion electrodes under defined mass transfer conditions. In such a cell, a vortex flow field is established where the fluid is circulating above a stationary electrode. In figure 3.2 the flow pattern of an ideal vortex is depicted. The fluid with a kinematic viscosity ν rotates inwards above an infinite stationary plane at a constant angular velocity ω. In the center of the vortex the fluid rises upwards. This flow regimen is very promising because it has been shown that an ideal vortex flow yields a constant hydrodynamic boundary layer thickness δ_V for a given rotation speed which is twice as thick as that of an RDE at the same rotation speed [79, 80]:

$$\delta_V = 8 \left(\frac{\nu}{\omega} \right)^{1/2} \tag{3.29}$$

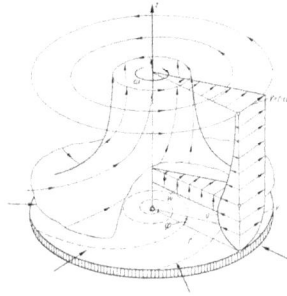

Figure 3.2: Ideal vortex flow field [78], with courtesy of Springer.

There are also other flow patterns which result in well-defined mass transfer to an electrode [81]. The RDE, the wall-tube, and wall-jet electrode yield well-defined and in case of the former two also homogeneous mass transfer. A wall-jet flow cell has been used successfully for studying CO oxidation kinetics [82, 83]. The cyclone flow cell employed in this study is comparatively easy to produce for DEMS experiments because there are no moving parts and the membrane can be sealed easily.

However, since the surface area in a cyclone cell is not infinite, cyclone walls are expected to have a significant influence on the flow. Sundmacher [78] analysed the cyclone flow in theory without taking into account wall friction and without analysing the thickness of the concentration boundary layer in detail. As discussed in the introduction, such information is important for analysing kinetics of porous electrodes quantitatively. To get a deeper insight into the flow pattern and especially the shape of the boundary layer, numerical flow simulations are performed.

In figure 3.3, the pathline of a single particle through the cell is illustrated. As expected, the fluid flows along the outer walls towards the bottom of the cell rising up in the center, producing a vortex above the membrane. There are some deviations from the ideal flow though: In an ideal vortex flow field the flow is axisymmetrical. The center of the cyclone flow does, however, not exactly correspond to the geometrical center of the cell which is ascribed to the positioning of the inlet. Classical cyclones for the separation of particles from a gas stream also show such a non-symmetric flow pattern [71]. Furthermore, for ideal vortex

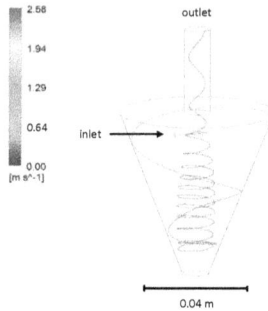

Figure 3.3: Simulated pathline and velocity of a single particle introduced at the inlet of the cyclone for a flow rate of 216 ml/min.

flow, the tangential velocity at the inner radius is not a function of the distance from the electrode. While the CFD results do not show a dependency of the tangential velocity on the distance from the electrode either (only in a small part of the cyclone near the inlet), the magnitudes of the tangential velocities from the CFD simulations are lower by a factor of 20 (100 ml/min) to 60 (590 ml/min) compared to the theoretical calculations in [78]. This is attributed to wall friction and fluid-fluid interaction between the flow in the center and the fluid descending along the outer walls which was not taken into account in the cited reference.

To obtain information about the concentration boundary layer under diffusion limited conditions, a surface reaction is introduced as explained in section 3.3.1. In figure 3.4, the thickness of the resulting concentration boundary layer, defined by $c_{boundary} = 0.9c_\infty$, for a flow rate of 590 ml/min for a mass transfer limited surface reaction is plotted over the electrode radius. It can be seen that the boundary layer has a relatively constant thickness between $R = 2.5 \cdot 10^{-3}$m and $R = 5 \cdot 10^{-3}$m. In the center where the fluid rises up, the boundary layer is much thicker. Thus only the outer part of the electrode features a homogeneous boundary layer. It should be noted, however, that this segment represents 75% of the total electrode area. The inner 25% of the electrode area feature a much thicker and less homogeneous boundary layer.

According to theory, the boundary layer thickness of a vortex flow should be proportional to angular velocity ω to the power of -0.5 (compare equation (3.29)). Sundmacher predicts a linear relationship between angular velocity, that is a

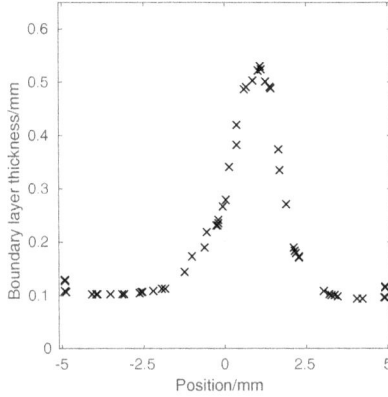

Figure 3.4: Simulated concentration boundary layer thickness as function of the distance from the electrode center at a flow rate of 216 ml/min.

function of the cross-radial velocity v_ϕ and the electrode radius R_{el}, and the inlet velocity which is directly correlated to the volume flow $\dot{V}$ via the inlet tube diameter d_{in} and the radius of the cyclone flow cell at the height of the inlet tube R_{in} [78]:

$$\omega = \frac{v_\phi}{R_{el}} = v_{inlet} \cdot \sqrt{\frac{R_{in}}{R_{el}}} \cdot \frac{1}{R_{el}} = \frac{4\dot{V}}{\pi d_{in}^2} \cdot \sqrt{\frac{R_{in}}{R_{el}}} \cdot \frac{1}{R_{el}} \quad (3.30)$$

Thus, it would be expected that since δ is proportional to $\omega^{-1/2}$ it would also be proportional to $\dot{V}^{-1/2}$. In figure 3.6, the thickness of the homogeneous part of the boundary layer $(R > 0.5R_{el})$ over the flow rate is depicted. Contrary to the prediction, a relationship of $\delta \propto \dot{V}^{-0.8}$ is obtained as the best fit. To rationalise this result, the dependency of angular velocity obtained from the simulation against flow rate is analysed. Angular velocity changes with the r and z-coordinate only very near the inlet tube, as explained above. Thus a direct correlation between flow rate and angular velocity can be obtained. It is found not not to be directly proportional to volume flow but to volume flow to the power of 1.68 as can be seen in figure 3.5. Combining these two findings the following relation is obtained:

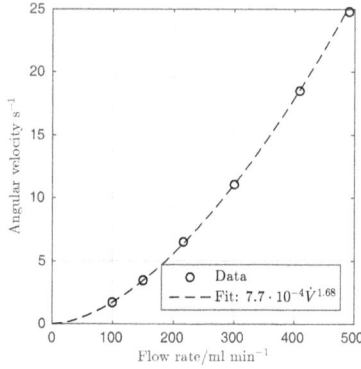

Figure 3.5: Simulated average angular velocity at $R = R_{el}$ for different flow rates.

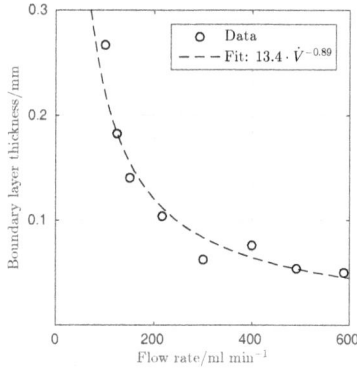

Figure 3.6: Simulated concentration boundary layer thickness for $R > 0.5R_{el}$ and different flow rates.

$$\omega \propto \dot{V}^{1.68} \rightarrow \dot{V} \propto \omega^{1/1.68} \tag{3.31}$$

$$\delta \propto \dot{V}^{-0.89} \rightarrow \delta \propto \omega^{-0.89/1.68} = \omega^{-0.52} \tag{3.32}$$

Thus the relationship $\delta \propto \omega^{1/2}$ is confirmed by the CFD simulation results whereas the dependence of ω on $\dot{V}$ seems to follow a different relation than

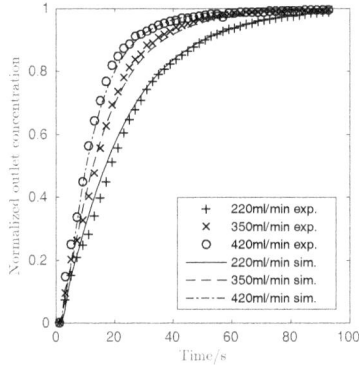

Figure 3.7: Simulated and measured residence time distribution curves at different flow rates.

predicted before - which is also supported by the deviation of the tangential velocity from the value calculated by Sundmacher [78] by an order of magnitude. For small cyclone flow cells, therefore, further studies should be conducted and the model by Sundmacher would need to be extended to cover non-ideal behaviour.

To validate the CFD calculations, residence time distributions were determined experimentally and simulated in Fluent. In figure 3.7, the experimental and simulated curves are shown. It can bee seen that the experimental residence time distributions are reproduced well by the simulations. Furthermore, variation of the diffusion coefficient in the simulation depicted in figure 3.8 leads to the well known behavior of $r_{\lim} \propto D^{2/3}$ resulting from laminar boundary theory [78, 70].

In DEMS experiments there is a further transport resistance because volatile species have to pass the porous PTFE membrane before entering the vacuum system. Therefore it is important to check if the transport of volatile species through the membrane into the vacuum is still proportional to the flow rate to the power of two-thirds. Here, mass transfer relations will be obtained from the flow analysis and expressed in terms of dimensionless numbers. Based on these numbers, the CFD results are compared with those from literature and experiments including the transport through the membrane. For this purpose, a solution saturated with CO_2 was pumped through the cell at different flow rates and the ion current was recorded by the mass spectrometer. The amount of CO_2 entering the vacuum can be determined using the calibration constant K^* of the

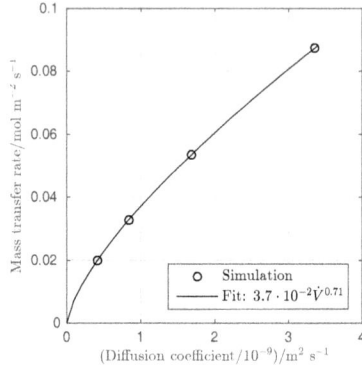

Figure 3.8: Simulated mass transfer rate over diffusion coefficient at a flow rate of 216 ml/min.

MS which will be discussed in the next section.

Sundmacher defined the Reynolds number for the cyclone flow cell as a function of electrode radius R_{el}, cyclone radius at inlet height R_{in}, diameter of the tangential inlet d_{in}, volume flow, and viscosity:

$$Re = \frac{4\dot{V}}{\pi d_{in}^2} \frac{\sqrt{R_{in} R_{el}}}{\nu} \tag{3.33}$$

and experimentally obtained the following mass transfer relation in terms of dimensionless numbers for the cyclone flow cell:

$$Sh = \frac{q}{1 - t_-} Re^{2/3} Sc^{1/3} \quad \text{(for } Re > 10^3 \text{ and } Sc \geq 1\text{)} \tag{3.34}$$

Sh denotes the Sherwood number, Sc the Schmidt number, t_- the anion transference number, and $q = 0.0136$ is a constant. In figure 3.9, the Sherwood numbers from experiment and simulation are depicted as a function of the Reynolds number as defined by equation (3.33). Mass transfer is proportional to flow rate to the power of two third in experiment and simulation, matching the relation obtained by Sundmacher. Thus the mass transfer to the vacuum in the DEMS cell can also be described by equation 3.34. Since the experimental data obtained by Sundmacher and in this work only covers Reynolds numbers between approximately 1000 and 35, 000, it is recommended to use the equations only with great care

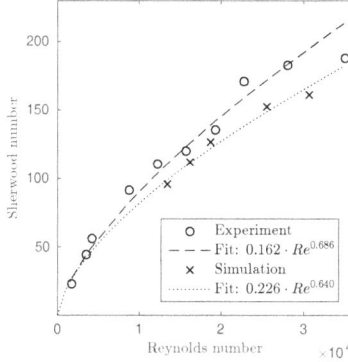

Figure 3.9: Sherwood number over Reynolds number from experiment and simulation, $\nu_{\mathrm{mixture}} = \nu_{H_2O} = 1.0410^{-6}\,\mathrm{m^2/s}$, $D_{\mathrm{A/B}} = 0.8410^{-9}\,\mathrm{m/s}$ [77], $D_{\mathrm{CO_2/H_2O}} = 1.9210^{-9}\,\mathrm{m/s}$ [77], solubility of CO_2 in water: $1.5\,\mathrm{g/kg}$ ($23\,^{\circ}\mathrm{C}$) [84].

outside this region. From equation 3.34 a Levich-type equation for the DEMS cyclone flow cell can be formulated.

The Levich equation, which can be expressed in dimensionless numbers as $Sh = 0.62Re^{1/2}Sc^{1/3}$, is usually applied in electrochemistry to calculate diffusion limited current densities i_{lim} in the following form:

$$i_{\mathrm{lim}} = 0.62z\mathrm{F}D^{2/3}\omega^{1/2}\nu^{-1/6}c_{\infty} \tag{3.35}$$

Accordingly, the mass transfer relationship for the constructed DEMS cyclone flow cell can be rewritten in the shape of a Levich-type relationship by re-substituting the dimensionless numbers in equation 3.34, yielding:

$$i_{\mathrm{lim}} = 0.021z\mathrm{F}D^{2/3}\left(\frac{4\dot{V}R_{\mathrm{in}}^{1/2}}{\pi d_{\mathrm{in}}^2 R_{\mathrm{el}}}\right)^{2/3}\nu^{-1/3}c_{\infty} \tag{3.36}$$

Although the Levich-type equation might be more familiar to some readers, equations 3.36 and 3.34 contain the same information. For the constant q in equation 3.34 a value of 0.021 is obtained from the simulation and 0.023 from the experiment. Experiment and simulation are thus in good agreement. The values do, however, differ slightly from the one obtained by Sundmacher.

Further investigation of the relationship between geometry parameters and the fluid velocity at the electrode and a review of the definition of the Reynolds number are necessary in order to be able to generalise equation 3.34 and the relation for the boundary layer thickness for cyclones of all kinds of shapes. Such an investigation might also resolve the difference in the constant q and the previous experimental results. For the purpose of conducting DEMS experiments under conditions of defined external mass transfer, the obtained relationship for the presented cell is sufficient, though.

3.4.2 DEMS calibration for CO_2

For quantitative analysis, calibration of the DEMS setup is necessary. Two separate calibration steps are performed: since not all molecules in the vacuum system get ionised and detected, the first calibration constant K^* describes the relationship between the amount of substance entering the vacuum system $\dot{n}_{MS}$ and the ion current I_{MS} at a fixed value m/z [42, 85, 86]:

$$K^* = \frac{I_{MS}}{\dot{n}_{MS}} \tag{3.37}$$

Since not all volatile species $\dot{n}_{reaction}$ which are produced in a reaction at the electrode enter the vacuum system, a second calibration constant K is defined. Besides K^* it takes into account the collection efficiency $0 \leq N = \frac{\dot{n}_{MS}}{\dot{n}_{reaction}} \leq 1$ of the cyclone cell which is influenced not only by the vacuum system and the MS itself but also by properties of the electrode and the membrane:

$$K = N \cdot K^* = \frac{I_{MS}}{\dot{n}_{reaction}} \tag{3.38}$$

In order to determine K^*, the vacuum system is connected through a leak valve to a chamber of known volume that can be filled with calibration gas. When monitoring the amount of gas leaking into the vacuum system and the ion current simultaneously, K^* can be determined.

Former studies already concluded that the electrochemical cell should be connected during calibration [42] because water diffusing through the membrane changes the total pressure and the ionization probabilities and thus influences the calibration constants for other substances.

The pressure in the calibration volume is generally very low because the calibration should preferably be performed at conditions close to the operating conditions. The temperature does not change significantly during the experiment because the

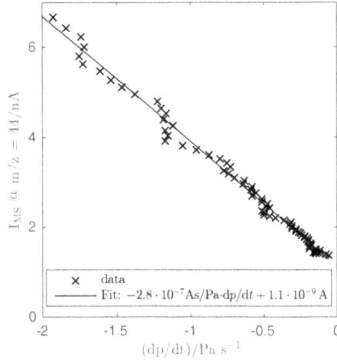

Figure 3.10: Ion current at m/z = 44 over the derivative of the pressure in the calibration volume while CO_2 is leaked into the vacuum system.

heat capacity of the metal walls enclosing the calibration volume is very large compared to that of the gas inside. Thus, the ideal gas law can be employed to calculate the amount of gas flowing into the vacuum system as a function of the pressure p inside the calibration volume:

$$\dot{n}_{MS} = -\frac{dp}{dt}\frac{V}{RT} \tag{3.39}$$

$$K^* = \frac{I_{MS}}{\dot{n}_{MS}} = -I_{MS}\frac{RT}{V}\left(\frac{dp}{dt}\right)^{-1} \tag{3.40}$$

Figure 3.10 shows the plot of ion current at m/z = 44 versus derivative of the pressure during calibration for CO_2. A linear correlation is observed and from the slope of the regression curve, $K^* = 0.186\,C/mol$ is calculated. Although it has been pointed out previously that the pressure measurement is a potential error source [85], this value was found to be very reproducible in the experiments with a standard deviation of 1.2% (n=3).

To obtain the DEMS calibration constant K for CO_2, the CO oxidation reaction is employed. Since CO_2 is the only reaction product, the amount of CO_2 produced at the electrode can be calculated from the Faradaic current I. K can be calculated from the following equation [42]:

$$K = N \cdot K^* = \frac{I_{MS}}{\dot{n}_{reaction}} = \frac{I_{MS} \cdot z \cdot F}{I} \tag{3.41}$$

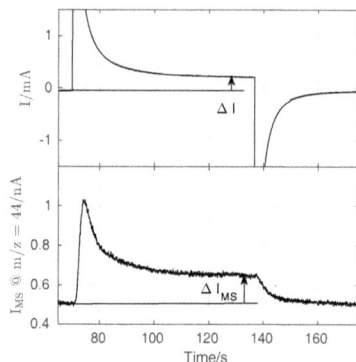

Figure 3.11: Faradaic current I and ion current I_{MS} over time for CO bulk oxidation. The potential is stepped from $0.45\,V$ to $0.65\,V$ at $t = 72\,s$ and back to $0.45\,V$ at $t = 137\,s$ at a flow rate of $430\,ml/min$ for the CO-saturated electrolyte containing $0.25\,mol/L$ $HClO_4$.

where $z = 2$ is the number of electrons transferred per molecule of CO_2.

CO-stripping during a CV is often used for calibration [42]. Since double layer contributions to the Faradaic current can be large especially for technical electrodes with high surface areas [87], steady state measurements at a constant electrolyte flow rate of $430\,ml/min$ are used for obtaining the calibration constant. For this purpose, the potential of the working electrode is stepped from a steady state at $0.45\,V$ to $0.65\,V$ and back again after reaching a second steadystate. In figure 3.11, the Faradaic and ion currents during a positive and a negative step are shown. The higher and lower potential are chosen based on cyclic voltammograms so that the lower potential is just above the onset potential of CO oxidation on Pt/Ru catalysts and the higher potential is well above the peak potential [87]. The steady state change in Faradaic current and ion current (ΔI_F and ΔI_{MS} in figure 3.11) is inserted into equation (3.41). This way, no additional background correction is required.

For CO_2, $K = 0.115\,C/mol$ is obtained. In four subsequent step experiments using the same electrode, a mean value of $0.117\,C/mol$ is obtained with a standard deviation of $6.7\,\%$. This corresponds to a collection efficiency of $N = 62\,\%$ for CO_2. The value of K is specific for the chosen electrode geometry, the electrolyte flow rate and composition, and the vacuum system / MS device. It can be expected that the collection efficiency drops at higher electrolyte flow rates because a larger share of the volatile products will diffuse into the electrolyte when the thickness of the diffusion layer decreases. Tegtmeyer and coworkers [49] present a very clear explanation for this phenomenon. In the next section an estimation of the collection efficiency of the cyclone cell will be discussed.

Obviously, this relation does not hold right after the potential step. The dynamic behavior will be discussed more detailed in section 3.4.3. In the same experiment, a calibration constant for CO can be obtained by monitoring the decrease of the CO signal instead of the increase of the CO_2 signal. The CO signal at $m/z = 28$ contains a comparatively large background signal from nitrogen. Moreover, CO_2 produces a signal at $m/z = 28$, too. The fact that different ions may contribute to the signal at a given m/z ratio has been discussed in literature before [88, 48, 89]. The intensity of the CO_2 signal at $m/z = 28$ was 18.4% of the intensity at $m/z = 44$ here. In literature, a value of 12.6% was reported for experiments conducted with a slightly higher ionization energy [90]. A calibration constant of $K = 0.09\,C/mol$ is obtained for CO after applying a correction for the contribution of CO_2 to the signal at $m/z = 28$. This seems reasonable, given the fact that the ionization cross section of CO_2 is higher than that of CO [91]. The calibration constant for CO_2 can be considered to be more reliable though.

In figure A.1 in the appendix, the flow of CO and CO_2 during the potential step experiment is depicted. A good quantitative agreement between the drop in CO and the increase in CO_2 flow is obtained.

3.4.3 Kinetics of CO oxidation at a porous electrode

In this section, kinetic parameters of the dynamic one-dimensional macrokinetic model for CO oxidation which was introduced in section 3.4.3 will be identified using the experimental data from the CO bulk oxidation experiment described in the previous section.

Overall, the assumption of a homogeneous boundary layer with only diffusive transport (compare section 3.3.2) seems reasonable for the given experimental conditions. The diffusion boundary layer has about the same thickness as the velocity boundary layer for a Schmidt number of one and is smaller for Schmidt number larger than one. Here the Schmidt number (ν/D) is about 500. Thus, the diffusion boundary layer is a lot thinner than the velocity boundary layer and the assumption that there is no convective transport is reasonable. The thickness

parameter	Value	Unit	
$k_{0,\text{OH,ad}}$	$0.0026 \cdot 10^{-5}$	s^{-1}	fitted
$k_{0,\text{OH,de}}$	$2.17 \cdot 10^{11}$	s^{-1}	fitted
$k_{0,\text{CO,ox}}$	$2.01 \cdot 10^{-6}$	s^{-1}	fitted
$k_{\text{CO,ad}}$	0.305	s^{-1}	fitted
N_{surface}	$4.09 \cdot 10^{16}$	-	fitted
g_{CO}	27.3	-	fitted
g_{OH}	13	-	[87]
α_{OH}	0.5	-	[87]
α_{CO}	0.5	-	[87]
β_{OH}	0.5	-	[87]
R	29.8	Ω	fitted
C_{dl}	0.0776	F	fitted

Table 3.1: Kinetic parameters for the CO oxidation modelling

of the diffusion layer is approximately 0.1 mm whereas the electrode diameter is 10 mm - since diffusive transport is proportional to the concentration gradient only a small mistake is introduced by neglecting the transport in other directions.

In total, eight parameters were fitted to the experimental data. A genetic algorithm starting with a set of randomly selected initial parameters is employed for finding the parameters that yield the best fit of experimental data and simulation results. The resulting parameter values can be found in table 3.1. In six runs the genetic algorithm, which started from a different set of randomly chosen parameter values each time, converged at very similar parameter values (standard deviation of 11 % on average). While this does not guarantee that there is only one set of parameters from a mathematical point of view, it shows that the fitting procedure is quite robust in spite of the high number of parameters. While the goal of this work was to establish the method, it is clear that a systematic experimental study would increase the reliability of the results of the CO oxidation kinetics.

In figures 3.12 and 3.13, the measured and simulated current as well as the measured and simulated flow of CO_2 into the vacuum system over time are depicted for a positive potential step from 0.45 V to 0.65 V at $t = 72$ s and a negative

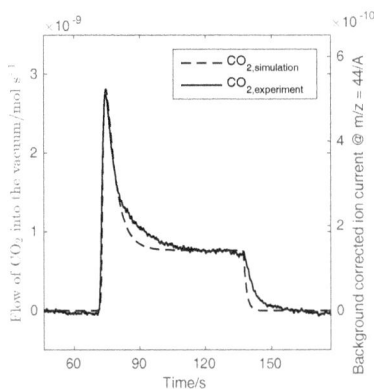

Figure 3.12: Simulated and measured CO_2 flow into the vacuum system over time during CO bulk oxidation. The potential is stepped from $0.45\,V$ to $0.65\,V$ at $t = 72\,s$ and back to $0.45\,V$ at $t = 137\,s$ at a flow rate of $430\,ml/min$ for the CO-saturated electrolyte containing $0.25\,mol/L$ $HClO_4$.

potential step back to $0.45\,V$ at $t = 137\,s$. The simulated flow of CO_2 from the membrane into the vacuum system is calculated according to equation 3.27. The measured flow of CO_2 is calculated from the background corrected ion-current signal at $m/z = 44$, which can be read directly on the right y-axis, multiplied with the calibration constant according to equation 3.40.

The current signal peaks almost instantly after the potential step whereas the CO_2 signal rises quickly and peaks after 4.2 seconds in both experiment and simulation. The relation between mass spectrometer signal and CO_2 reaction rate that can be expressed as the calibration constant K for steady state conditions does not hold then because of transport effects and double layer charging. Subsequently, the curve falls and stabilises at approximately $7 \cdot 10^{-10}\,mol/s$. A small but steady decay of current and CO_2 flow into the vacuum can be observed in the experimental curves. This behavior has been observed before for platinum based catalysts and has been attributed to anion adsorption [92]. After the positive potential step the MS-signal increases quickly whereas it drops slightly slower after the negative potential step. The slow decay is mainly caused by diffusion of carbon dioxide in electrolyte, catalyst, and membrane. Overall there is a close agreement of the simulation results and the experimental data.

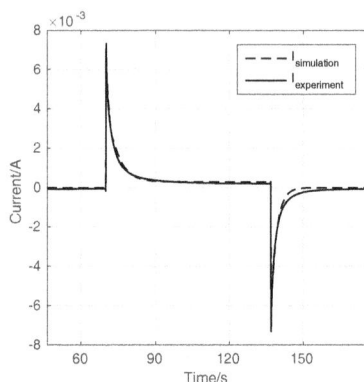

Figure 3.13: Simulated and measured current over time during CO bulk oxidation. The potential is stepped from $0.45\,V$ to $0.65\,V$ at $t = 72\,s$ and back to $0.45\,V$ at $t = 137\,s$ at a flow rate of $430\,ml/min$ for the CO-saturated electrolyte containing $0.25\,mol/L$ $HClO_4$.

From the simulation results, not only kinetic parameters can be estimated. The concentration profiles and local reaction rates can be accessed and give further insight into the interaction of transport and reaction processes: In figure 3.14, the steady state concentration profile of CO and CO_2 in the boundary layer, the electrode and the membrane is depicted as a function of the z-coordinate for high and low potential. It can be seen that at low potential the concentration of CO is declining from the bulk value at the outer boundary layer to zero at the membrane-vacuum interface while there is no CO_2 present. During the oxidation at a potential of $0.65\,V$, the CO_2 that is produced in the electrode is diffusing through the membrane into the vacuum and through the boundary layer into the bulk where the concentration approaches zero. The CO concentration drops over the electrode, leading to the local reaction rate being only half as large at the membrane side of the electrode as on the electrolyte side.

Because of the initially high CO concentration and the high CO surface coverage in the electrode, the reaction rate increases sharply and exceeds the steady state value after a positive potential step. This is causing the CO_2 signal to overshoot as depicted in figure 3.12. The diffusion through the membrane causes a slight delay of the CO_2 signal compared to the current signal. On the other hand, after a negative potential step, the CO_2 which diffuses from the concentration

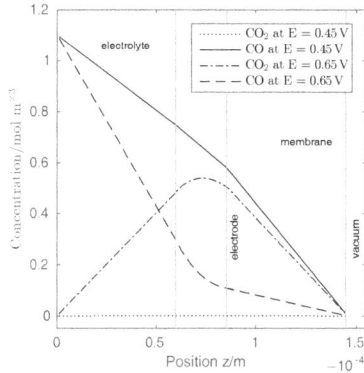

Figure 3.14: Simulated steady state concentration profile of CO and CO_2 in the boundary layer (left), the electrode (middle) and the membrane (right) over the position at 0.45 and 0.65 V.

boundary layer and the electrode through the membrane into the vacuum leads to a delayed decay of the amount of CO_2 entering the vacuum. There is a slight difference between simulation and experiment after the negative potential step. Three phenomena contribute to this difference: Firstly, the limited speed of the vacuum pumps might cause some tailing of the measured signal. Secondly, the model underestimates the thickness of the concentration boundary layer in the middle of the electrode. From there CO_2 produced at high potential might diffuse back to the electrode for a longer period of time than predicted by the model. Finally, local CO_2 concentrations might exceed the solubility during CO oxidation at high potential, causing transport phenomena which were not included in the model.

Additionally, the collection efficiency of the cyclone cell in steady state was estimated at different flow rates by changing the thickness of the diffusion layer in the model and calculating the collection efficiency N: Increasing the flow rate to 600 ml/min leads to a predicted value of N of 0.5. Decreasing the flow rate to 300 ml/ min results in a collection efficiency of approximately 0.75, decreasing it further to 200 ml/min in $N = 0.85$.

In conclusion, the model is capable of giving an insight into the dynamic transport and reaction processes and their interaction inside a porous DEMS electrode. Parameter identification using the online CO_2 measurement from the

DEMS in addition to current and potential has advantages over using current and potential only. Effects such as double layer capacity and anion adsorption are difficult to separate from the reaction currents by monitoring current and potential alone but they do not influence the direct online detection of reaction products by DEMS.

3.5 Conclusions

A new cyclone flow DEMS cell for analysing products of electrochemical reactions inside porous electrodes online has been presented. The designed cyclone cell features a homogeneous concentration profile over 75% of the electrode area, a simple design without any moving parts, defined convection and high collection efficiency. The disadvantages of classical DEMS cells such as badly defined mass transfer and locally varying concentrations are circumvented. Though the CFD simulations of the mass transfer are sufficient for the purpose of the current study it would certainly be interesting to resolve the deviation to the literature values which were discussed. It needs to be pointed out that the comparatively high flow rates required by the cyclone flow cell are a disadvantage because they restrict the use of expensive reactants and make long term experiments difficult. The second point is less severe considering the fact that the cell is designed for examining fast processes rather than for undertaking long term studies. A methodology how to combine DEMS experiments, flow analysis and physical modelling to gain quantitative understanding of processes inside porous electrodes was demonstrated. The model based description of the electrode processes allows reliable parameter identification and simultaneous analysis of electrode transport properties and electrochemical properties. While the well-researched example of CO oxidation served as an example to establish the methodology, it will be shown in the following chapters that this set up and approach is useful for analysing kinetics of different kinds of reactions on porous electrodes.

Chapter 4

Methanol oxidation on a Pt/Ru catalyst

Having established the dynamic concentration measurements by DEMS, the combination of dynamic DEMS measurements and physical simulations is extended towards the more complex methanol electrooxidation reaction (MOR) in this chapter. The dynamics of the MOR are expected to occur in a wider range of time scales than those of CO oxidation. Thus EIS and CV are applied for the MOR because they allow to cover dynamic processes in a wide range of time scales with high accuracy.

4.1 Introduction

MOR kinetics on various catalysts have already been discussed intensely in literature for model electrodes [93, 94, 95] as well as technical electrodes [1, 3, 96]. In the latter three references, mathematical models have been developed that describe current potential relationships for a carbon supported Pt/Ru catalyst under steady state and dynamic conditions. This type of catalyst is also employed here because of its relevance for application in fuel cells. The electrochemical oxidation reaction is described by the following steps:

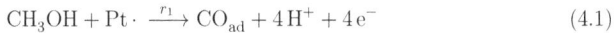

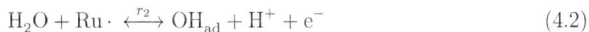

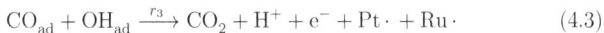

$$CH_3OH + Pt \cdot \xrightarrow{r_1} CO_{ad} + 4\,H^+ + 4\,e^- \tag{4.1}$$

$$H_2O + Ru \cdot \xleftrightarrow{r_2} OH_{ad} + H^+ + e^- \tag{4.2}$$

$$CO_{ad} + OH_{ad} \xrightarrow{r_3} CO_2 + H^+ + e^- + Pt \cdot + Ru \cdot \tag{4.3}$$

The stable adsorbate on the surface is CO which further reacts with adsorbed OH to CO_2. It is assumed that OH only adsorbs on free Ru sites (Ru·) and CO only on free Pt sites (Pt·).

The interaction of the three reaction steps, mass transfer effects, dynamically changing surface coverages, and double layer charging make the MOR on technical electrodes a complex system. EIS is a powerful technique that allows to separate and investigate phenomena on a wide range of time scales and was already shown to be a suitable tool for investigating MOR kinetics [96]. However, processes with similar time constants cannot be distinguished in EIS spectra. To separate and analyse such processes, an additional measurement signal can be used.

A number of techniques described in literature can be used to analyse the relationship of sinusoidal signals of concentrations and electrical quantities: Hartmann and co-workers [97, 98] observed the pressure response of a closed metal-air battery that contained gaseous oxygen to analyse the transport of the reactant in the cell and established the term electrochemical pressure impedance spectroscopy (EPIS). Engebretsen et al. [99] used the same term for a slightly different approach. They set the pressure inside a PEMFC as an input signal and monitored the electrical response. Another approach used for the investigation of a PEMFC is concentration-alternating frequency response analysis (cFRA) [100]. In cFRA, the response of current and voltage to perturbations of the oxygen partial pressure were analysed to investigate mass transfer and cathode humidification.

Based on these promising results from literature, we propose to combine frequency response analysis of species flux and electrochemical impedance spectroscopy as a tool to study the dynamics of transport processes and to separate the influences of different species and reactions. Because of low response times and possibility of quantification DEMS is a suitable tool for measuring species fluxes for the purpose of FRA.

DEMS experiments have been used to quantify products of the MOR such as carbon dioxide and methylformate at different potentials in mass spectrometric cyclic voltammograms (MSCV) [101, 55, 58] or to systematically study current efficiency and product composition for alloys of Pt_x/Ru_{1-x} with x between one and zero [102]. While DEMS has been used successfully to obtain qualitative and quantitative data on MOR reaction mechanism and kinetics, DEMS data has not yet been correlated quantitatively to mechanistic models of the MOR macrokinetics. DEMS measurements under dynamic conditions other than MSCVs include step changes in current density [103] and step changes in methanol inlet concentration [65]. Here DEMS and EIS will be combined for the first time for FRA.

In this chapter, it is explored whether the combination of electrical and species flux-based FRA allows to separate processes that are connected to mass transfer from the reaction processes. Accordingly, the focus and novelty of the work presented in this chapter lies in the methodology rather than on gaining new insights into the well-studied MOR reaction mechanism.

First, the concept of species flux-based FRA will be explained and basic equations will be developed. Next, the experimental set up and measurement procedures for the DEMS measurements of the MOR as well as the mathematical model that describes the MOR kinetics and the transport of CO_2 into the MS will be described. The model is based on the work by Krewer et al. [96]. In the following section, the identification of a single parameter set that describes current and CO_2 production during CV and EIS will be shown. The parameter set is validated by comparing experimental data and simulation results. Factors influencing the species flux-based FRA such as amplitude, frequency range and influence of the membrane transport into the vacuum are discussed. The chapter closes with a brief conclusion which focuses on the advantages and limitations of the new technique.

4.2 Theory of species flux-based frequency response analysis

In EIS, the complex impedance value $Z(\omega) = \frac{\Delta U}{\Delta I}$ contains direct information about all processes that effect electrical current and potential such as reaction kinetics, capacitive effects and ohmic resistance. Mass transfer effects only have an indirect effect on the impedance because changes in concentration affect the reaction kinetics.

DEMS allows to measure fluxes of volatile substances directly and with a time resolution that is high enough to record the response to a sinusoidal input signal. Thus, a relationship between current and mass transfer of volatile species can be established. Only the linear part of the response signal is considered and the following transfer function $G_{MS}(\omega)$ is defined:

$$G_{MS}(\omega) = \frac{\Delta I_{MS}}{\Delta I} \tag{4.4}$$

This value can be directly obtained from experimental data by dividing the signals of input current I and MS ion current I_{MS} at a given mass to charge ratio in the frequency domain. The same mathematical transformation procedures that are used to convert current and voltage signals from the time domain into the frequency domain can be used. In theory it is also possible to replace the current by the potential, which results in a different transfer function. However, the representation from equation 4.4 seems more practical for the purpose of separating reaction and transport phenomena because reaction rates and mass transfer are directly related to current and only indirectly related to potential. Since EIS is already a well established tool to examine current-potential relations, the focus is placed on the current-species flux relation here.

For better comparability between measurement devices and different reactions, a normalization of the transfer function is undertaken. The MS signal is converted into the molar flux that enters the vacuum $\dot{n}_{MS}$ using the MS calibration constant K^* that can be determined as explained in 3.4.2. The current signal is converted into a theoretical molar production rate $\dot{n}_I$ that is calculated by Faraday's law with the number of electrons per molecule of product z.

$$G_{MS,n}(\omega) = \frac{\Delta I_{MS}}{K^*} \frac{zF}{\Delta I} = \frac{\Delta \dot{n}_{MS}}{\Delta \dot{n}_I} \tag{4.5}$$

The normalised transfer function $G_{MS,n}$ is thus in theory independent of electrode area, number of transferred electrons per molecule of the measured compound, and sensitivity of the MS. It can approach a limiting value of one at low frequencies if no side reactions occur and if all product molecules enter the vacuum system.

If the magnitude of $G_{MS,n}$ is very small, a large part of the current is used for double layer charging, production of intermediates or side reactions in the respective frequency range. If $G_{MS,n}$ is larger than one, dynamic processes amplify the change in production rate so that $\Delta \dot{n}_{MS} \geq \Delta I zF$. At a limiting frequency of $0\,Hz$ and without any side products, the magnitude of $G_{MS,n}$ shows the share of product that enters the vacuum system and equals the DEMS collection efficiency N that was defined in section 3.4.2.

Further characteristic properties of the transfer function will be discussed in section 4.7 specifically for the MOR.

The explanations given above are valid if the anodic current direction is defined as positive and if the compound measured by DEMS is a product of an electrochemical oxidation reaction, which is the case for CO_2 in the MOR. If a reaction educt is measured, the production rates become consumption rates and the sign of the calculated transfer function would change when using the same equations.

4.3 Experimental

The same cell and DEMS set up as in chapter 3 was used to study the MOR and record FRA spectra. The fluid flow inside the cell as well as the collection efficiency are discussed in detail there. The electrodes were produced in the same manner and using the same catalyst and binder as described above. $5\,\mathrm{g\,m^{-2}}$ of catalyst were applied, corresponding to a layer thickness of $25\,\mu m$ determined by SEM measurements. The electrolyte that was circulated through the cell at a constant flow rate of $230\,\mathrm{ml\,min^{-1}}$ contained $0.5\,\mathrm{mol\,L^{-1}}$ methanol (VWR, HPLC grade) and $0.25\,\mathrm{mol\,L^{-1}}$ $HClO_4$ (Sigma Aldrich, ACS grade). No further methanol was added during the experiments because from the current density it was calculated that methanol consumption is negligible over the duration of the experiment. All measurements were carried out at a constant room temperature of $298\,\mathrm{K}$.

The MS was calibrated using the same procedure as explained in section 3.4.2 and a calibration constant of $0.0347\,\mathrm{C\,mol^{-1}}$ was obtained.

Two types of dynamic electrochemical experiments were carried out: cyclic voltammetry and electrochemical impedance spectrometry. During all experiments current, voltage and the mass spectrometer's ion current signal at a mass to charge ratio m/z of 44 were recorded to detect the expected main product CO_2. Because the ion current was recorded in parallel to the electric current during electrochemical impedance measurements, the potentiostat's inbuilt EIS function could no be used. Instead, sinusoidal input current signals were applied to the working electrode. The responses of potential and ion current were recorded in the time domain and transformed to the frequency domain by a fast fourier transformation using a LABVIEW program that was developed specifically for this purpose [104].

EIS and flux-based FRA spectra were recorded in parallel with an amplitude of $50\,\mathrm{mV}$ and an DC offset of $0.7\,\mathrm{V}$ in the frequency range from 0.02 - 1 Hz. The reason for the choice and the effect of the comparatively high amplitude on the linearity of the system are discussed below. CVs were recorded between $0.45\,\mathrm{V}$

and 0.8 V. The second cycle is shown. Below 0.45 V there is only small CO_2 production and above 0.8 V ruthenium dissolves from the catalyst. Like in the previous chapter, all potentials are reported with respect to a reversible hydrogen electrode.

4.4 Modelling

In this section, a model for the MOR is developed to analyse the measurement data and discuss the FRA results in detail. The interpretation of classical EIS data is often not straightforward because the impact of processes can overlap in the same frequency range and the time constants of individual processes are unknown. However, various examples from literature have demonstrated the advantages of employing physical models to interpret EIS spectra [1, 2, 105]. It is expected that a physical model will be essential for the interpretation of species flux-based FRA because no examples of interpretations are available in literature.

The description of the reaction kinetics will be based on a previously published non-linear physical model of a direct methanol fuel cell by Krewer et al. [96]. Transport is described similarly as in chapter 3 because the same DEMS cell is used. A preliminary version of the model presented in this section was developed within a master thesis by Bo Yuan [106].

The structure of the DEMS model for the MOR is depicted in figure 4.1.

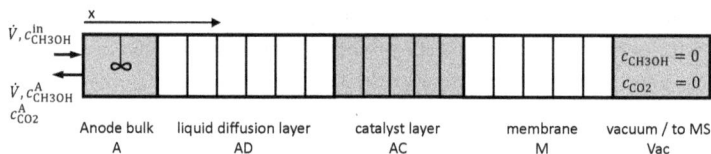

Figure 4.1: Schematic overview of the structure of the DEMS model for the MOR

Methanol diffuses from the cell bulk volume through the liquid diffusion layer into the catalyst layer where the reactions take place. Additionally methanol can evaporate through the porous PTFE membrane. CO_2 produced in the catalyst layer can either diffuse through the membrane into the vacuum system or through the diffusion layer in to the bulk.

It has been shown that the changing surface coverages of adsorbed CO and OH have a significant influence on the MOR reaction kinetics. Thus the following balance equations for adsorbed OH and CO species are included, with $\theta_{CO/OH}$ denoting the relative surface coverage of the catalyst sites and $c_{Pt/Ru}$ denoting the number of available surface sites on the catalyst surface per unit of geometrical electrode surface area:

$$\frac{d\theta_{CO}}{dt} = \frac{1}{c_{Pt}}(r_1 - r_3) \tag{4.6}$$

$$\frac{d\theta_{OH}}{dt} = \frac{1}{c_{Ru}}(r_2 - r_3) \tag{4.7}$$

The rates r_1, r_2 and r_3 of the three individual reaction steps 4.1, 4.2 and 4.3 are calculated with a Frumkin/Temkin adsorption mechanism:

$$r_1 = k_{10}\exp[-\beta_{CO}g_{CO}(\theta_{CO} - 0.5)]c_{CH_3OH}^{AC}(1 - \theta_{CO}) \tag{4.8}$$

$$r_2 = k_{20,f}\exp\left[\frac{\alpha F}{RT}\eta\right]\exp\left[-\beta_{OH}g_{OH}(\theta_{OH} - 0.5)\right](1 - \theta_{OH})$$
$$- k_{20,b}\exp\left[-\frac{(1-\alpha)F}{RT}\eta\right]\exp[(1 - \beta_{OH})g_{OH}(\theta_{OH} - 0.5)]\theta_{OH} \tag{4.9}$$

$$r_3 = k_{30}\exp[(1 - \beta_{CO})g_{CO}(\theta_{CO} - 0.5)]\theta_{CO}\theta_{OH} \tag{4.10}$$

$g_{OH/CO}$ is the inhomogeneity/interaction factor for Frumkin/Temkin adsorption, α is the charge transfer coefficient, k_{10}, $k_{20,f}$, $k_{20,b}$, and k_{30} are the reaction rate constants of the three reactions, and β_{CO} and β_{OH} is the symmetry parameter for Frumkin/Temkin adsorption.

The reaction overpotential η is calculated from a charge balance with the double layer capacity C_{dl} and the external cell current density i_{cell}. The sum of the overpotential and the reaction's equilibrium potential E^0 equals the electrode potential E. The external potential $E_{external}$ that can be measured experimentally includes the uncompensated ohmic electrolyte resistance R_u:

$$E_A = \eta + E^0 \tag{4.11}$$

$$\frac{d\eta}{dt} = \frac{1}{C_{dl}}i_{cell} + \frac{1}{C_{dl}}(-4Fr_1 - Fr_2 - Fr_3) \tag{4.12}$$

$$E_{external} = E + R_u i_{cell} \tag{4.13}$$

The following balance equations describe the transport of methanol in the system:

$$\frac{\mathrm{d}\, c^{\mathrm{A}}_{CH_3OH}}{\mathrm{d}\, t} = \frac{\dot{V}}{V^{\mathrm{A}}} \left(c^{\mathrm{A,in}}_{CH_3OH} - c^{\mathrm{A}}_{CH_3OH} \right) + \frac{A_{\mathrm{el}}}{V^{\mathrm{A}}} D_{CH_3OH} \left. \frac{\partial c^{\mathrm{DL}}_{CH_3OH}}{\partial x} \right|_{\mathrm{A/DL}} \tag{4.14}$$

$$\frac{\mathrm{d}\, c^{\mathrm{DL}}_{CH_3OH}}{\mathrm{d}\, t} = -D_{CH_3OH} \frac{\partial^2 c^{\mathrm{DL}}_{CH_3OH}}{\partial x^2} \tag{4.15}$$

$$\frac{\mathrm{d}\, c^{AC}_{CH_3OH}}{\mathrm{d}\, t} = -D^{\mathrm{AC}}_{CH_3OH} \frac{\partial^2 c^{\mathrm{AD}}_{CH_3OH}}{\partial x^2} - \frac{r_1}{\delta^{\mathrm{AC}}} \tag{4.16}$$

$$\frac{\mathrm{d}\, c^{\mathrm{M}}_{CH_3OH}}{\mathrm{d}\, t} = -D^{\mathrm{M}}_{CH_3OH} \frac{\partial^2 c^{\mathrm{M}}_{CH_3OH}}{\partial x^2} \tag{4.17}$$

The superscripts A, DL, AC and M denote the respective domain as indicated in figure 4.1 with A/DL, DL/AC, and AC/M being the interfaces between A and DL, DL and AC, and AC and M. $\dot{V}$ is the volumetric flow rate, $c^{\mathrm{A,in}}_{CH3OH}$ is the methanol feed concentration, V^{A} the anode chamber volume, A^{el} is the electrode surface area. $D^{\mathrm{M}}_{CH_3OH}$ the diffusion coefficient of methanol through the membrane and $D^{\mathrm{AC}}_{CH_3OH}$ is the diffusion coefficient of methanol in the anode catalyst layer. δ^{AC} is the thickness of the anode catalyst layer.

The thickness of the diffusion layer in front of the catalyst layer δ^{DL} is obtained from figure 3.6 in chapter 3.

Transport of CO_2 is crucial for the DEMS model. CO_2 that is produced in the anode reaction can either diffuse back into the bulk electrolyte or through the PTFE membrane into the vacuum system. The following balance equations are established to describe the transport of CO_2:

$$\frac{\mathrm{d}\, c^{\mathrm{A}}_{CO_2}}{\mathrm{d}\, t} = -\frac{\dot{V}}{V^{\mathrm{A}}} c^{\mathrm{A}}_{CO_2} + \frac{A_{\mathrm{el}}}{V^{\mathrm{A}}} D_{CO_2} \left. \frac{\partial c^{\mathrm{DL}}_{CO_2}}{\partial x} \right|_{\mathrm{A/DL}} \tag{4.18}$$

$$\frac{\mathrm{d}\, c^{\mathrm{DL}}_{CO_2}}{\mathrm{d}\, t} = -D_{CO_2} \frac{\partial^2 c^{\mathrm{DL}}_{CO_2}}{\partial x^2} \tag{4.19}$$

$$\frac{\mathrm{d}\, c^{\mathrm{AC}}_{CO_2}}{\mathrm{d}\, t} = -D^{\mathrm{AC}}_{CO_2} \frac{\partial^2 c^{\mathrm{DL}}_{CO_2}}{\partial x^2} + \frac{r_3}{\delta^{\mathrm{AC}}} \tag{4.20}$$

$$\frac{\mathrm{d}\, c^{\mathrm{M}}_{CO_2}}{\mathrm{d}\, t} = -D^{\mathrm{M}}_{CO_2} \frac{\partial^2 c^{\mathrm{M}}_{CO_2}}{\partial x^2} \tag{4.21}$$

Continuity of the CO_2 and methanol flux is ensured at the boundaries between anode, anode diffusion layer, catalyst layer and membrane with the following boundary conditions:

$$-D_{CH_3OH} \left. \frac{\partial c_{CH_3OH}^{DL}}{\partial x} \right|_{DL/AC} = -D_{CH_3OH}^{AC} \left. \frac{\partial c_{CH_3OH}^{AC}}{\partial x} \right|_{DL/AC} \tag{4.22}$$

$$-D_{CO_2} \left. \frac{\partial c_{CO_2}^{DL}}{\partial x} \right|_{DL/AC} = -D_{CO_2}^{AC} \left. \frac{\partial c_{CO_2}^{AC}}{\partial x} \right|_{DL/AC} \tag{4.23}$$

$$-D_{CH_3OH}^{AC} \left. \frac{\partial c_{CH_3OH}^{AC}}{\partial x} \right|_{AC/M} = -D_{CH_3OH}^{M} \left. \frac{\partial c_{CH_3OH}^{M}}{\partial x} \right|_{AC/M} \tag{4.24}$$

$$-D_{CO_2}^{AC} \left. \frac{\partial c_{CO_2}^{AC}}{\partial x} \right|_{AC/M} = -D_{CO_2}^{M} \left. \frac{\partial c_{CO_2}^{M}}{\partial x} \right|_{AC/M} \tag{4.25}$$

$$\tag{4.26}$$

At the membrane/vacuum interface, the CO_2 and methanol concentrations are set to zero because the partial pressures in the vacuum can be neglected.

The flux of CO_2 into the vacuum $\dot{n}_{MS}$ is calculated from the concentration gradient at the interface M/Vac between membrane and vacuum:

$$\dot{n}_{MS} = -A_{el} D_{CO_2} \left. \frac{\partial c_{CO_2}^{M}}{\partial x} \right|_{M/Vac} \tag{4.27}$$

Diffusion coefficients of CO_2 and CH_3OH in the porous catalyst layer are calculated via the Bruggemann equation:

$$D_j^{AC} = D_j (\varepsilon^{AC})^{1.5} \tag{4.28}$$

Diffusion coefficients of methanol and CO_2 in the membrane are identified from experimental data. This approach is chosen because the representation of transport through the membrane as diffusive transport is a simplification of the physical process. At the vacuum side of the membrane, where the pressure in the pores is very low, transport might not be diffusive any more. Exploring the membrane transport in detail is, however, out of scope of this work.

The values of all fixed model parameters are reported in table 4.1. The equations were implemented in Matlab. Transport equations were solved with a finite volume scheme using $N_{elements}^{DL} = 5$ volume elements for the diffusion layer, $N_{elements}^{AC} = 5$ volume elements for the catalyst layer, and $N_{elements}^{M} = 5$ volume elements for the

Table 4.1: Fixed model parameters

Parameter	Value	Notes
δ^{DL} /m	$100 \cdot 10^{-6}$	see section 3.4.1
δ^{M} /m	$60 \cdot 10^{-6}$	SEM measurement
δ^{AC} /m	$25 \cdot 10^{-6}$	see section 3.2.4
ϵ^{AC} /-	0.92	see section 3.2.4
ϵ^{M} /-	0.72	see section 3.2.4
$\dot{V}$ /L s^{-1}	$3.67 \cdot 10^{-3}$	from experiment
A_{el} /m^2	$78.5 \cdot 10^{-6}$	from experiment
c_{CH3OH}^{in} /mol m^{-3}	500	from experiment
T /K	298	from experiment
E^0 /V	-0.036	Nernst equation with standard potential from [95]
D_{CH3OH} /m^2 s^{-1}	$0.84 \cdot 10^{-9}$	[77]
D_{CO2} /m^2 s^{-1}	$1.92 \cdot 10^{-9}$	[77]
g_{OH} /-	0.43	[96]
g_{CO} /-	11	[96]
β_{OH} /-	0.5	[96]
β_{CO} /-	0.5	[96]
α /-	0.5	[96]
$N_{elements}^{DL}$ /-	5	
$N_{elements}^{AC}$ /-	5	
$N_{elements}^{M}$ /-	5	

membrane. This comparatively low number of elements is chosen to limit the computational cost of the parameter identification that will be explained in the next section. The impact of the number of volume elements on the simulation results is discussed in the results section. Simulation results were converted from the time domain into the frequency domain using a fast Fourier transform.

4.5 Parameter identification

The parameters for the MOR model from Krewer et al. [96] were determined for a temperature of 343 K whereas experiments in this work were carried out at a temperature of 296 K. Furthermore the catalyst particles may be different. Therefore, kinetic rate constants from [96] cannot be used. Also, as explained above, diffusion coefficients in the porous PTFE membrane are treated as additional fitting parameters.

The model parameters θ are identified by minimizing the deviation between experiment and simulation result using the following objective function f:

$$\min_{\theta} \quad f = \epsilon_{\mathrm{CV}} + \epsilon_{\mathrm{FRA}} \tag{4.29}$$

$$\epsilon_{\mathrm{CV}} = \frac{W_{\mathrm{CV,i}}}{n_{\mathrm{CV}}} \sum_{i=1}^{n_{\mathrm{CV}}} \left(I_i^{\mathrm{exp}} - I_i^{\mathrm{sim}} \right)^2$$

$$+ \frac{W_{\mathrm{CV,MS}}}{n_{\mathrm{CV}}} \sum_{j=1}^{n_{\mathrm{CV}}} \left(I_{\mathrm{MS},j}^{\mathrm{exp}} - I_{\mathrm{MS},j}^{\mathrm{sim}} \right)^2 \tag{4.30}$$

$$\epsilon_{\mathrm{FRA}} = \frac{W_{\mathrm{FRA,MS}}}{n_{\mathrm{FRA}}} \sum_{i=1}^{n_{\mathrm{FRA}}} \left(\mathrm{Im}(G_{\mathrm{MS},i}^{\mathrm{exp}}) - \mathrm{Im}(G_{\mathrm{MS},i}^{\mathrm{sim}}) \right)^2 + \left(\mathrm{Re}(G_{\mathrm{MS},i}^{\mathrm{exp}}) - \mathrm{Re}(G_{\mathrm{MS},i}^{\mathrm{sim}}) \right)^2$$

$$+ \frac{W_{\mathrm{FRA,i}}}{n_{\mathrm{FRA}}} \sum_{i=1}^{n_{\mathrm{FRA}}} \left(\mathrm{Im}(Z_i^{\mathrm{exp}}) - \mathrm{Im}(Z_i^{\mathrm{sim}}) \right)^2 + \left(\mathrm{Re}(Z_i^{\mathrm{exp}}) - \mathrm{Re}(Z_i^{\mathrm{sim}}) \right)^2 \tag{4.31}$$

With ϵ_{FRA}, ϵ_{CV} denoting the deviation between simulated and experimental Nyquist and CV curves. ϵ_{FRA} and ϵ_{CV} are calculated by summing up the squared differences between experimental and simulated mass spectrometric and electrochemical data.

For the CV, the squared deviations between experiment and simulation of current i_i and ion current $I_{\mathrm{MS},i}$ are directly summed up over the n_{CV} data points that were recorded. The weight factors $W_{\mathrm{CV,i}}$, $W_{\mathrm{CV,MS}}$, $W_{\mathrm{FRA,i}}$ and $W_{\mathrm{FRA,MS}}$ take into account that the ion current and current signals differ by several orders of magnitude (mA vs. nA) and allow to place different weight on the two experiments.

For the species flux-based FRA and EIS spectra, the deviation between experiment and simulation is calculated from the squared distances between the experimental and the simulated Nyquist curve that are summed up over the number of frequencies n_{FRA}.

The objective function is minimised by the genetic algorithm from Matlab's global optimization toolbox because gradient-based algorithms were only able to find local optima of the objective function. The following values were used for the weight factors: $W_{CV,i} = 4.86 \cdot 10^5 \, A^{-2}$, $W_{CV,MS} = 5.6 \cdot 10^{19} \, A^{-2}$, $W_{FRA,i} = 0.12 \, \Omega^{-2}$ and $W_{FRA,MS} = 2.68 \cdot 10^{15}$. The weights were obtained by trial and error and the quality of the fit was evaluated visually. In section 6.5.3, it will be shown how parameters can be identified from a set of experimental data with different measured quantities without chosen weight factors by using a log likelihood function as objective function.

The genetic algorithm typically needs about 80.000 evaluations of the objective function before converging which is reasonable giving the number of parameters. Since every evaluation of the objective function includes the simulation of an mass spectrometric cyclic voltammogram (MSCV) and a simulation of the response to a sinusoidal input signal for each frequency, a computer with 8GB of RAM and an i7 Quad-Core CPU needs about 15 hours for the parameter identification even though a single simulation only takes one to two seconds, depending on the parameters.

4.6 Model validation

In figure 4.2, the experimental and the simulated CV and MSCV are compared. The overall shape of the curves matches well. Experimental current density and CO_2 flux, which is calculated from the ion current signal and the MS calibration constant, are well reproduced by the model with the identified parameter set. Because of the low scan rate the CV does not show many characteristic features for CVs. The main deviation between experiment and simulation results from the fact that the experimental values of current density and CO_2 flux do not return to their starting value at the end of the cycle. The reason for this might be a shift of the ion current background. While the data can be used to demonstrate the methodology, the parameter values should be treated with caution. In figure 4.3, the experimental and the simulated EIS and flux-based FRA Nyquist plots are compared. Only the first harmonic, which corresponds to the linear part of the response signal, is analysed here. The Nyquist plot of the EIS shows one large distorted semicircle that comprises the reactions as well as diffusion and double

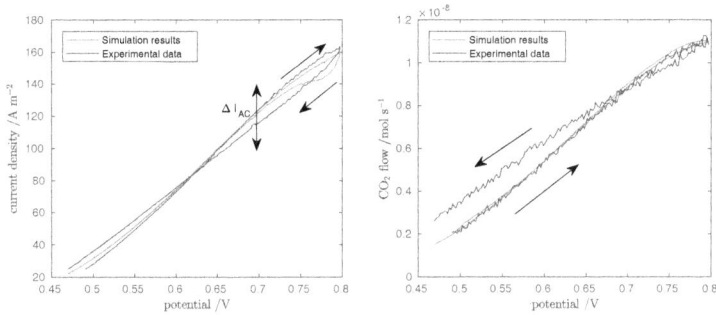

Figure 4.2: Comparison of experiment and simulation for CV and MSCV. Scan rate $2\,\mathrm{mV/s}$, $c_{\mathrm{MeOH}} = 0.5$ mol/L in $0.1\,\mathrm{mol/L}$ $HClO_4$, $0.5\,\mathrm{mg/cm^2}$ Pt/Ru on carbon, T=25° C.

layer charging. The experimental data show some noise at higher frequencies which might be caused by the use of the in-house Labview program that does not contain elaborate noise filtering and is not optimised for data acquisition speed. The experimental Nyquist plot from the flux-based FRA curve shows two characteristic features which are both reproduced by the simulation. First, the imaginary part of the transfer function is positive at high frequencies. Second, the real part of the transfer function is negative at high to intermediate frequencies. The reason for this behaviour will be discussed in section 4.7.

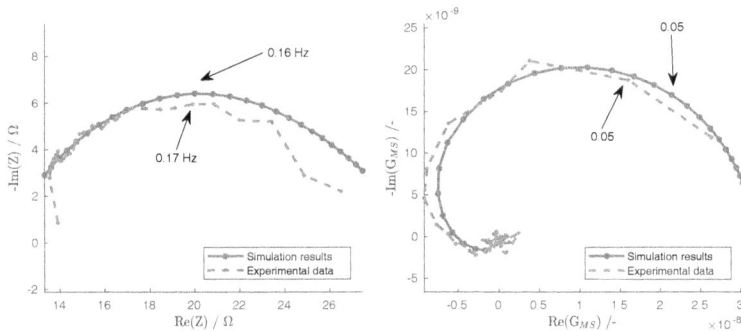

Figure 4.3: Comparison of experiment and simulation for EIS and flux-based FRA. Frequency: 0.02 - 1 Hz, $c_{\mathrm{MeOH}} = 0.5\,\mathrm{mol/L}$ in $0.1\,\mathrm{mol/L}$ $HClO_4$, $0.5\,\mathrm{mg/cm^2}$ Pt/Ru on carbon, T=25° C.

Parameter	Best fit	Krewer et al. [96]
k_{10} /m s^{-1}	$8.12 \cdot 10^{-8}$	$1.6 \cdot 10^{-4}$
k_{20f} /mol m^{-2} s^{-1}	$1.43 \cdot 10^{-6}$	$7.2 \cdot 10^{-4}$
r_{20b} /mol m^{-2} s^{-1}	$1.59 \cdot 10^{7}$	$9.91 \cdot 10^{-4}$
r_{30} /mol m^{-2} s^{-1}	8.07	0.19
C_{dl} /F m^{-2}	226.6	3348
(F g$_{catalyst}^{-1}$)	(55.3)	(41.9)
R_{ohm} /Ω	12.3	-
c_{Pt} /mol m^{-2}	$2.65 \cdot 10^{-3}$	0.117
(mol g$_{catalyst}^{-1}$)	$(0.53 \cdot 10^{-3})$	$(2.06 \cdot 10^{-3})$
c_{Ru} /mol m^{-2}	0.018	0.165
(mol g$_{catalyst}^{-1}$)	$(3.6 \cdot 10^{-3})$	$(1.46 \cdot 10^{-3})$
D_{CH3OH}^{M} /m^2 s^{-1}	$5.70 \cdot 10^{-10}$	$6.26 \cdot 10^{-10}$ [a]
D_{CO2}^{M} /m^2 s^{-1}	$1.26 \cdot 10^{-9}$	-

^{a} = in Nafion.

Table 4.2: Comparison of identified parameter values for MOR on Pt/Ru and literature values.

All together, the simulation model reproduces CV, MSCV, EIS, and flux-based FRA spectra comparatively well with a single parameter set. In table 4.2, the values of the identified model parameters and the parameter values from Krewer et al. [96] are shown. It can be seen that the identified kinetic parameter values differ strongly from the values found in [96]. The reasons for the deviation might be the difference between half cell and full cell set up, the different amounts of catalyst and its preparation, the difference between liquid and solid electrolyte, or the different temperature. The double layer capacitance and numbers of surface sites show large differences when normalised by the geometric surface area of the electrode. However, when accounting for the fact that the catalyst loading in [96] was 16 times larger than in this study, the numbers of surface sites per gram of catalyst are in the same order of magnitude and the double layer capacitance agrees well with literature values. The value of the diffusion coefficient of CO_2 in the membrane is very close to the value calculated via the Bruggeman equation from the membrane porosity and the diffusion coefficient of CO_2 in the electrolyte of $1.17 \cdot 10^{-9}$ m^2 s^{-1} that was used in chapter 3. Also, the value of the diffusion coefficient of methanol in the membrane agrees well with the value calculated via the Bruggeman equation from the diffusion coefficient of methanol in the electrolyte of $5.13 \cdot 10^{-9}$ m^2 s^{-1}. Thus the identification of the diffusion coefficients might not be necessary and literature values might be used in future studies.

4.7 Results and discussion

In this section the flux-based FRA spectra are analysed in detail using insights from the simulation. The MSCV is not analysed in detail because flux-based FRA as a novel technique is the focus of this chapter. The surface coverages of CO and OH during the MSCV can be found in figure A.2 in section A.2 of the appendix. First, technical aspects are covered. It will be examined if the comparatively small experimental frequency range from 0.02 to 1 Hz is suitable for investigating the MOR reaction. Also, the influence of the comparatively large excitation amplitude will be analysed and the influence of the sensitivity and background noise of the mass spectrometer as well as the excitation amplitude on the frequency range that can be measured will be evaluated systematically.

Next, the electrochemical and transport processes of the MOR on Pt/Ru in the DEMS cell will be analysed using the FRA spectrum. The origin of the negative real and imaginary part of the spectrum will be explained and assigned to specific processes. Finally, a sensitivity analysis will show which parameters influence the spectra and may thus be experimentally accessible through the technique.

4.7.1 FRA frequency range

The experimental frequency range from 0.02 to 1 Hz is very narrow compared to EIS where frequencies between 1 mHz and 1 MHz are used. Thus a wider frequency range is simulated using the validated model to investigate if any important features of the curve have been missed out. Figure 4.4 shows the EIS and flux-based FRA Nyquist plot for an extended frequency range of 0.0003-10 Hz. Here, the normalised transfer function $G_{MS,n}$ is used. The experimental frequency range and some frequency points are highlighted for clarity. It can be seen that the curve does not show any significant features outside the experimental frequency range but quickly converges to zero at high frequency and intercepts with the real axis at low frequency. The EIS Nyquist plot approaches the real axis for higher and lower frequencies. Thus the experimental frequency range is appropriate for the MOR in the current set up.

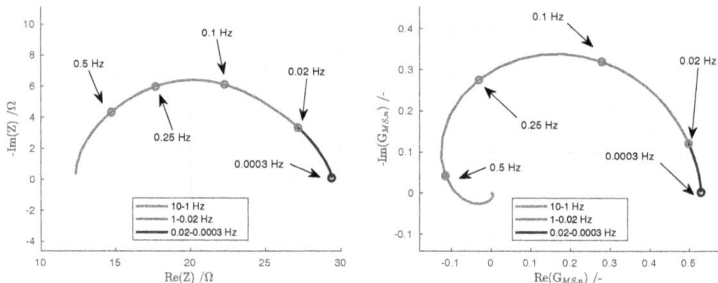

Figure 4.4: Simulated EIS (left) and flux-based FRA (right) Nyquist diagram in an extended frequency range of 0.0003-10 Hz showing that the characteristic features of the curves are within the experimental frequency range of 0.02 - 1 Hz.

4.7.2 Non-linearity at high input current magnitude

The magnitude $|Z|$ of the EIS impedance $Z = \frac{\Delta U}{\Delta I}$ changes with the frequency. This means that the amplitude of the AC current perturbation also changes with the frequency during the potentiostatic EIS experiments. At 1 Hz the amplitude of the current signal is 3.6 mA, and at 0.02 Hz it is 1.8 mA. In the following, a constant current amplitude will be used for the simulations because the flux-based FRA is defined as a function of the ion current and the additional frequency dependency of the amplitude would make analysis more complex.

Figure 4.5 shows the flux-based FRA Nyquist plot simulated with three different input current amplitudes: 0.1, 2 and 3 mA. From the deviation between the curves at 0.1 and 3 mA it is evident that the amplitude of 3 mA, that corresponds to the experimental conditions, already causes some non-linearities. These non-linearities are not a problem when comparing the experimental results to a simulation model that also contains non-linear effects as is the case here. They would, however, impede the comparison of experimental data and analytical expressions that rely on the assumption of small perturbations. In the next paragraph it will be shown that the experimental amplitude can only be reduced at the cost of the frequency range.

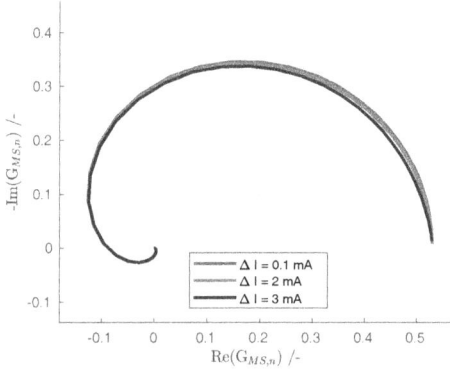

Figure 4.5: Simulated flux-based FRA Nyquist diagram in an extended frequency range from 0.0003-10 Hz for input current amplitudes between 0.1 and 3 mA showing that an influence of non-linearities is present but not large at the experimental input current amplitudes.

4.7.3 Relationship between MS signal noise and maximum frequency

While the simulated flux-based FRA curves are smooth, the experimental MS signal data contains noise. The noise becomes more severe at higher frequencies. This effect can be seen in figure 4.3 where the experimental data points are masked by strong noise at frequencies above 0.5 Hz. The background noise of the MS is constant. Thus, at higher frequencies, where the magnitude of the transfer function and thus the absolute amplitude of the MS signal is smaller, a constant noise level is relatively more severe. When the noise level exceeds the MS signal's amplitude, the signal does not contain sufficient information and cannot be transformed into the frequency domain correctly and random data points are produced by the fast Fourier transformation.

Thus, the following condition must be met:

$$|\Delta I_{\text{MS}}| = |G_{\text{MS}}||\Delta I| \overset{!}{>} P_{\text{N,MS}} \qquad (4.32)$$

$$|G_{\text{MS,min}}| \overset{!}{>} \frac{P_{\text{N,MS}}}{|\Delta I|} \qquad (4.33)$$

The magnitude of the transfer function $|G_{\text{MS}}|$ must be larger than the ion current noise $P_{\text{N,MS}}$ divided by the ion current signal amplitude.

From the MS signal in the time domain, a noise level of approximately $20\,\text{pA}$ is determined. In figure 4.6 (left), $|G_{\text{MS,min}}|$ is depicted in the flux-based FRA Nyquist plot for two different values of ΔI. For all points within the circle, the current I_{MS} corresponding to the amplitude of the CO_2 flux into the vacuum is smaller than the noise of the MS signal. Thus, to record the most important features of the spectrum, the high amplitude is required.

Since the magnitude of the FRA transfer function $|G_{\text{MS}}|$ is a function of the frequency ω, the maximum frequency can be predicted for given input current amplitudes and MS background noise levels with the simulation model. In figure 4.6 (right), the maximum attainable frequency is plotted over the amplitude of the current input signal for three different noise levels: $10\,\text{pA}$, $20\,\text{pA}$, and $40\,\text{pA}$. The maximum frequency increases with input signal amplitude in all cases. The shape of the curve is convex, meaning that the benefits of higher input signals diminish at higher input signal magnitudes.

The predicted values agree reasonably well with the experimental values. According to the simulations, the maximum frequency for a meaningful spectrum is $0.6\,\text{Hz}$ at a current amplitude of $3\,\text{mA}$. The maximum frequency in the experiment is approximately $0.45\,\text{Hz}$ at a current amplitude of $3.2\,\text{mA}$.

The input signal magnitude in figures 4.5 to 4.6 are reported in absolute values. This representation was chosen because the signal to noise ratio of the MS depends on the absolute species flow rather than on the area specific values. Furthermore, the electrode area cannot be scaled easily in the DEMS set up since the vacuum pumps have a limited capacity. However, the current amplitude can be normalised by the electrode area of $0.78\,\text{cm}^2$ if desired.

4.7.4 Contributions of transport and reaction processes

In the following the contributions of the processes in the DEMS cell to the flux-based FRA spectrum will be discussed.

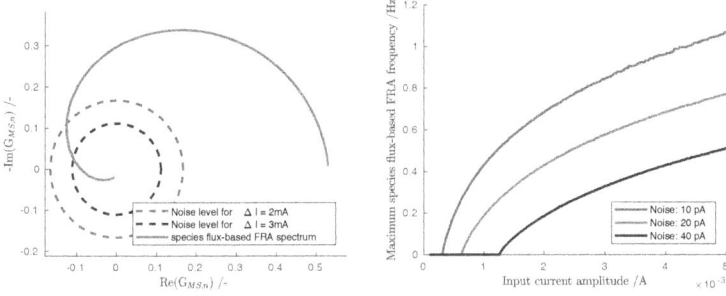

Figure 4.6: Left: flux-based FRA Nyquist plot with noise circles for different input current signal amplitudes in a frequency range of 0.002-10 Hz. Right: Calculated maximum frequency for flux-based FRA over input current signal magnitude for different MS signal noise levels. The blue line corresponds to the experimental conditions. The curves are not smooth at high input current amplitudes because the spectra are computed for a finite number of discrete frequencies.

At high frequencies, the flux-based FRA Nyquist plot starts at the origin, evolves clockwise in a spiral shape, and finally converges towards the real axis at low frequencies. In order to explain the shape of the Nyquist plot and gain a better understanding of the processes that determine the shape, the transfer function is split up into the contributions of relevant processes:

$$G_{\mathrm{MS,n}} = \frac{\Delta \dot{n}_{\mathrm{MS}}}{\Delta \dot{n}_{\mathrm{I}}} = \underbrace{\frac{\Delta \dot{n}_{\mathrm{R}}}{\Delta \dot{n}_{\mathrm{I}}}}_{G_{\mathrm{R,n}}} \cdot \underbrace{\frac{\Delta \dot{n}_{\mathrm{M}}^{\mathrm{in}}}{\Delta \dot{n}_{\mathrm{R}}}}_{G_{\mathrm{RM,n}}} \cdot \underbrace{\frac{\Delta \dot{n}_{\mathrm{MS}}}{\Delta \dot{n}_{\mathrm{M}}^{\mathrm{in}}}}_{G_{\mathrm{M}}} \tag{4.34}$$

$G_{\mathrm{M}} = \frac{\Delta \dot{n}_{\mathrm{MS}}}{\Delta \dot{n}_{\mathrm{M}}^{\mathrm{in}}}$ is the transfer function for the transport through the membrane into the vacuum. $\Delta \dot{n}_{\mathrm{MS}}$ is the flux of CO_2 into the vacuum and $\Delta \dot{n}_{\mathrm{M}}^{\mathrm{in}}$ is the flux of CO_2 that enters the membrane from the catalyst layer. $G_{\mathrm{R,n}} = \frac{\Delta \dot{n}_{\mathrm{R}}}{\Delta \dot{n}_{\mathrm{I}}}$ is the transfer function for the CO_2 producing reaction. $\Delta \dot{n}_{\mathrm{R}} = \int_0^{\delta^{\mathrm{AC}}} \frac{r_3}{\delta^{\mathrm{AC}}} \, dx$ is the rate of CO_2 production. $\Delta \dot{n}_{\mathrm{I}}$ is the theoretical rate of CO_2 production calculated from the total current by Faraday's law. $G_{\mathrm{RM,n}} = \frac{\Delta \dot{n}_{\mathrm{M}}^{\mathrm{in}}}{\Delta \dot{n}_{\mathrm{R}}}$ is the transfer function for the transport from the reaction sites in the catalyst layer into the membrane.

In figure 4.7, the reaction transfer function $G_{\mathrm{R,n}}$ is depicted. The curve starts at the origin of the coordinate system at high frequencies. The reason for this is the fact that at high frequencies the input current entirely goes into double layer charging and discharging so that no change in the CO_2 production is induced and the magnitude of the transfer function is zero. At low frequencies, the curve approaches the real axis and the magnitude becomes one. In this range, dynamic effects can be neglected and the CO_2 production equals the value predicted by Faraday's law. This reflects the fact that the model does not account for any side reactions.

Additionally, the magnitude of the transfer function is larger than one in an intermediate frequency range. This means that the CO_2 production increases and decreases stronger with the current than predicted by Faraday's law. The reason for this is the interplay of surface processes: During the positive half-wave of the current input signal, the surface coverage of adsorbed intermediates (CO and OH) is decreased, causing a disproportionally large increase in CO_2 production. Subsequently, the surface coverage is built up again during the negative half-wave causing a disproportionally large decrease in CO_2 production rate.

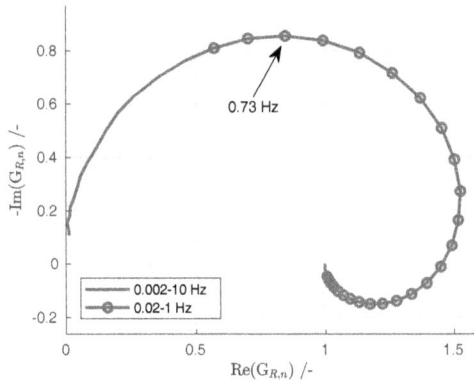

Figure 4.7: Simulated flux-based FRA Nyquist plot of the electrode reaction only ($G_{R,N} = \Delta n_{CO2}^{prod}/\Delta n_I$) in the frequency range of 0.002-10 Hz. The experimental frequency range of 0.02-1 Hz is indicated with open circles.

In figure 4.8, the simulated membrane transfer function is shown. $G_{M,n}$ has a magnitude of one at low frequencies meaning that the flow of CO_2 into the membrane equals the flow out of the membrane which is an important condition for obeying the law of mass conservation. The membrane transport causes negative real and imaginary values at higher frequencies. In the time domain (not shown) it can be seen that unlike in EIS the phase shift increases continuously with increasing frequency. When the phase shift exceeds $\pi/2$, a negative real part results. Such behaviour may occur for systems with more than one differential equation that have more than one pole in the transfer function.

The number of differential equations, i.e. discretisation elements of the membrane $N = 5$, is comparatively low, considering the importance of the transport of CO_2 through the membrane for the flux-based FRA spectrum. Therefore, simulations with three different element numbers were carried out. The results for the membrane transfer function are shown in figure 4.8. It can be seen that the discretisation of the membrane does not have an effect on the qualitative features of the curve. The chosen number of five elements is a compromise between simulation time and accuracy. As described in section 4.5, the global optimiser limits the acceptable computational complexity of the model because thousands of spectra are simulated during the parameter identification process.

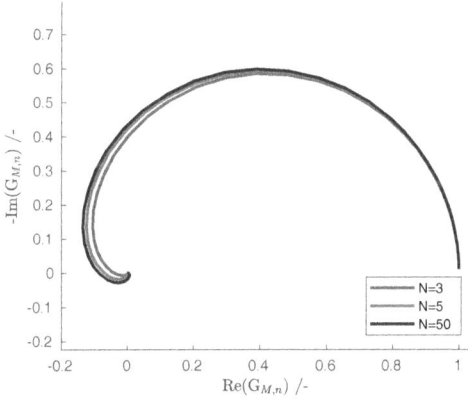

Figure 4.8: Simulated Nyquist plot of the PTFE membrane only ($G_{\mathrm{M}} = \frac{\Delta \dot{n}_{\mathrm{MS}}}{\Delta \dot{n}_{\mathrm{M}}^{\mathrm{In}}}$) with different numbers of finite volume elements N in a frequency range of 0.002-10 Hz.

The Nyquist plot of the transfer function of the transport from the reaction sites to the membrane $G_{\mathrm{RM,n}}$ is depicted in figure 4.9. The magnitude at low frequency is different from one because only part of the CO_2 that is produced diffuses to the membrane while the rest diffuses into the bulk electrolyte. Thus the magnitude of the overall flux-based FRA transfer function at low frequencies is determined by this process. It does, however, not cause a large phase shift, as is evident from the small imaginary part. The two arcs are produced by the interaction of the two transport processes into the membrane and into the bulk electrolyte. However, since these processes were not resolved in detail in the model and since they have a small influence on the phase shift, they are not discussed in detail here.

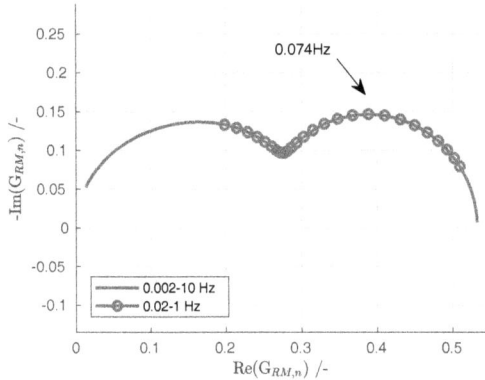

Figure 4.9: Simulated Nyquist plot of the transport from the catalyst layer to the membrane ($G_{RM,n} = \frac{\Delta \dot{n}_M^{in}}{\Delta \dot{n}_R}$) in a frequency range of 0.002-10 Hz. The experimental frequency range of 0.02-1 Hz is indicated with open circles.

The overall transfer function and the transfer functions of all processes discussed above are summarised in figure 4.10. The reaction shows a special feature with the magnitude of the normalised transfer function exceeding the value of one in an intermediate frequency range. However, this feature is not visible in the overall spectrum. The reason for this is that various processes overlap in the intermediate frequency region. Only at the high and low end of the frequency range, features can easily be assigned to individual processes: The negative real part of the overall spectrum at high frequencies is caused by transport through the membrane. The behaviour at the lowest frequencies is exclusively determined by the transport from the catalyst into the membrane since the magnitude of the transfer functions of all other processes are one there. At slightly higher frequencies, the inductive influence of the reaction and the capacitive influence of the membrane compensate each other.

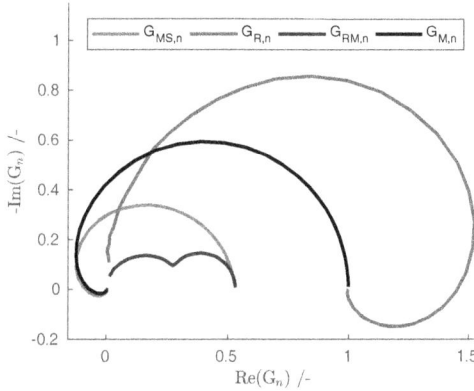

Figure 4.10: Decomposition of the flux-based FRA Nyquist curve into the transfer functions of individual processes in a frequency range of 0.002-10 Hz.

4.7.5 Sensitivity analysis

To investigate which physical parameters affect the flux-based FRA Nyquist curve, two relative sensitivity measures S_R and S_f are used as suggested in [98]:

$$S_R = \frac{R/R_0}{P/P_0} \tag{4.35}$$

$$S_f = \frac{f_r/f_{r,0}}{P/P_0} \tag{4.36}$$

f_r is the characteristic frequency where the imaginary part of the transfer function reaches its minimum. R is the difference in the real part of the transfer function between the lowest and the highest frequency. Since the real part of the transfer function approaches zero at high frequencies and the imaginary part approaches zero at low frequencies, R generally equals the magnitude of the transfer function at the lowest frequency. P_0 is the nominal value of a parameter. P is set 5 % higher than P_0. Accordingly $f_{r,0}$ and R_0 refer to the characteristic frequency and magnitude of the transfer function for the nominal parameter value P_0.

Thus S_R describes the relative sensitivity of the magnitude at lowest frequency towards a parameter value P, whereas S_f describes the relative sensitivity of the phase shift towards a parameter. While these two values do no capture the complete shape of the curves, they allow to quantify and compare the effect of different parameters on the flux-based FRA Nyquist curve in terms of frequency and magnitude.

In figure 4.11, the relative sensitivities for selected geometric, transport and reaction parameters are shown. The thickness of the membrane δ^M and the

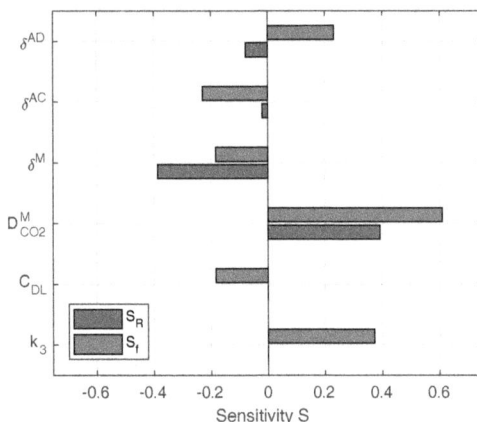

Figure 4.11: Sensitivities of flux-based FRA Nyquist spectra on selected geometric, transport and reaction parameters

diffusion coefficient of CO_2 in the membrane D_{CO2}^M have an opposite effect on the magnitude R at the lowest frequency. If the membrane is thicker, a smaller share of the CO_2 enters vacuum and if the diffusion is faster, a larger share enters the vacuum. Double layer capacity, C_{DL}, and rate constant k_3 do not have any influence on R because it is only influenced by the collection efficiency of the DEMS for a system without side reactions. The characteristic frequency f_r is influenced by all parameters, including the thickness of the catalyst layer δ^{AC}. This finding agrees with the conclusions from the previous section because f_r lies in an intermediate frequency range near $0.1\,\text{Hz}$ that is influenced by all processes.

In summary, the flux-based FRA spectra are sensitive to a number of parameters that often have opposing effects. Especially transport through the PTFE membrane, which is usually not of primary interest in electrochemical systems, has a strong influence. Both factors pose a challenge when it comes to separating different processes.

4.8 Conclusions

In this chapter, dynamic DEMS experiments and physical modelling were combined to investigate the MOR. A model for the MOR on a porous Pt/Ru catalyst was presented. A single parameter set was identified from experimental data that could be used to describe current, potential and MS data during CV, EIS and flux-based FRA. The existing literature model from Krewer et al. [96] could thus be verified. Some of the identified parameter values differ significantly from the literature values, though.

Furthermore, with flux-based FRA a new dynamic experimental technique was introduced and its application was demonstrated for the MOR. The interpretation of spectra and the contribution of different transport and reaction processes as well as the sensitivities of flux-based FRA transfer function towards various parameters were discussed in detail.

While the experimental data could be reproduced with the model, the quality of the CV data is is not excellent because the MS background signal was drifting over time. During FRA measurements, the signal was stable though. Thus the methodology has a lot of potential but the kinetic constants have to be interpreted with great caution.

The interpretation of flux-based FRA spectra is still at the very beginning, since this is the first time this methodology is applied. Still some general conclusions can be drawn from the analyses presented above:

- The frequency range of flux-based FRA is limited at the upper end by the noise level of the MS signal and at the lower end by the time constants of the processes of interest.

- The maximum frequency can be increased when using larger current signal amplitudes because this increases the absolute amplitude of the MS signal. At high amplitudes, the benefits of increasing the amplitude further diminish.

- The current amplitude that is necessary for a frequency range up to $0.5\,Hz$ already causes non-linear effects. This is acceptable when using a non-linear physical model for the interpretation of the spectrum. For analysing the spectra without a physical model, linear behaviour is important though. Thus, the amplitude of the current signal cannot be increased arbitrarily and the characteristics of the electrochemical cell should be carefully considered when choosing the current signal amplitude. In this context potentiostatic operation (modulation of the potential) is beneficial: it causes higher current amplitudes at high frequencies and thus increases the upper frequency range. At the same time, the current amplitudes are lower at lower frequencies where non-linear effects might occur.

- A principle limitation of flux-based FRA results from the fact that the diffusive transport processes lie in a narrow range of time constants compared to processes that affect EIS where time constants of ohmic drop, double layer charging, and reactions can differ by several orders of magnitude. This makes the separation of reaction and transport processes difficult. The sensitivity analysis presented above confirms this finding. In this work, modelling helped to provide additional insights into the contributions of individual processes to the spectra. Physical models and reaction kinetics are not available for many electrochemical reactions, though.

Considering the limitations explained above, it seems likely that steady state measurements of reaction products together with EIS and other purely electrochemical techniques would in most cases yield similar information as flux-based FRA in its current state of development. However, future work might elucidate how to obtain additional insights utilizing the frequency resolved species flux data.

A way of overcoming one limitation and increasing the flux-based FRA frequency range without introducing higher non-linearities might be to use high amplitudes in the high frequency region where there are hardly any non-linear effects and to use lower amplitudes at lower frequencies, where non-linear effects are more pronounced. This strategy requires, however, a good understanding of the dynamics of the system.

Despite its limitations, flux-based FRA has one key advantage over other techniques: The MS can differentiate between species. In EIS only electrons are measured and in EPIS the sum of all gaseous species is measured. A promising application where this advantage of flux-based FRA may be utilised in future is the electrochemical CO_2 reduction because the educt as well as volatile products such as CH_4 and H_2 might be distinguished.

Part 2 – Bioelectrochemical oxidation reactions

Chapter 5

Acetate oxidation in a biofilm electrode[1]

In part one, dynamic DEMS experiments and modelling have been described for two electrochemical reactions. Parameters for mathematical models were identified from the measurements and a good agreement between simulation and experiment was reached. The reaction mechanisms and model equations for CO and methanol oxidation were taken from literature. In the following, the methodology will be extended towards the more complex bioelectrochemical acetate oxidation in a BES for which electron transfer mechanisms are not clear yet.

5.1 Introduction

Gaining deeper knowledge on electron transfer mechanisms and energy conversion in BES is not only of scientific interest but also highly relevant for system design for technical applications. There are numerous *in-situ* and *ex-situ* techniques for studying anode respiring bacteria (ARB) [107] but most of them focus on the transfer of electrons out of the cell to an electrode. Analysis of metabolic processes in the cell are usually not conducted online but rather *ex-situ* [108]. These techniques also require to destroy the biofilm irreversibly. One of the most popular dynamic techniques for studying phenomena related to extra-cellular electron transfer between a biofilm and an electrode is cyclic voltammetry. As described in chapter 2, two types of CVs are distinguished in the field of BES [26]: non-turnover and turnover CVs. Non-turnover CVs are recorded using substrate-depleted solutions, i.e. at conditions where there is no substrate available to the biofilm. The current recorded then is caused exclusively by capacitive effects and the oxidation and reduction of redox systems within the cell. When no

---

[1]Parts of this chapter have been published F. Kubannek, U. Schröder, U. Krewer, Revealing metabolic storage processes in electrode respiring bacteria by differential electrochemical mass spectrometry, Bioelectrochemistry 121 (2018) 160–168.

substrate oxidation or other irreversible processes take place, the integrals of positive and negative currents over time are equal and the average current is zero. As no substrate oxidation takes place under non-turnover conditions, assigning oxidation and reduction of these redox systems to processes observed in the substrate metabolising biofilm needs to be done with uttermost care. In contrast, turnover CVs are recorded at high substrate availability. Since substrate oxidation is irreversible, the average current under turnover-conditions is usually positive. Current peaks from oxidation and reduction of inter-cellular redox systems are masked by the large current from substrate oxidation. Thus, it is difficult to clearly identify the role that each redox system plays in substrate oxidation from turnover as well as from non-turnover CVs.

Direct non-invasive measurement of metabolites or products such as CO_2 during CVs would help to understand the contributions of the redox systems that are visible in CVs to anodic respiration. So far, measurements of gas production in microbial fuel cells or electrolysis cells have been limited to headspace gas analysis by gas chromatography and quantification by measuring the total volume of collected gas [109, 110, 111]. The reported approaches did not achieve the time resolution needed to monitor product formation during CV measurements. One reason is that the equilibrium reaction between hydrogen carbonate and CO_2 slows down the transfer of CO_2 from a buffered solution near pH 7 into the gas phase [112] so that even small headspace volumes and intense stirring are unlikely to solve this problem. An alternative technique applied for non-electrochemical reactions is membrane inlet mass spectrometry (MIMS). With this technique, volatile products and intermediates from normal bacteria have been analysed online [113, 114]. MIMS yields a high time resolution and sensitivity because volatile substances pervaporate through a gas-permeable membrane into a vacuum system where they are detected by mass spectrometry. It has already been shown that CO_2 production in bioreactors can be monitored and quantified this way [115, 112]. Following these studies, the use of DEMS in combination with static and dynamic electrochemical techniques for the study of metabolic processes of electrochemically active bacteria is a promising approach. While DEMS has been numerously used to study electrode reactions [42, 43], so far no bioelectrochemical reactions have been investigated by DEMS. In this chapter, DEMS is combined with cyclic voltammetry of electroactive anodic biofilms for the first time. MSCV that allow to directly correlate CO_2 production and current at high and low substrate availability are reported, and electron transfer systems that are related to CO_2 production are identified. It is demonstrated that DEMS can be employed for quantitative analysis of the CO_2 production rate of electrochemically active bacteria and two intra-cellular storage mechanisms for charge and substrate as well as their implications on the analysis of BES are discussed.

5.2 Experimental

5.2.1 Electrochemical cell

The experiments were conducted in the cyclone flow DEMS cell which was described and characterised in detail earlier (see [116] and chapter 3). In brief, a working electrode made from PTFE-free carbon paper (r=0.5 cm, d = 0.2 mm, Sigracet GDL 29AA) is mounted at the bottom of a cyclone flow cell that allows to establish defined mass transfer conditions at the electrode. A porous PTFE membrane (Pall Membranes, specified pore size 0.2 μm, thickness 60 μm) is pressed against the bottom of the working electrode separating the working electrode compartment from the mass spectrometer's vacuum system. Through this membrane, volatile substances evaporate into the vacuum where they are detected by a mass spectrometer (Pfeiffer QMG220 M1 quadropole mass spectrometer with secondary ion multiplier). Because of the direct contact of electrode and membrane, volatile species have a short diffusion path into the vacuum system. The ion current at a mass to charge ratio of m/z = 44 (CO_2) was recorded throughout the duration of the experiment. The complete setup is depicted in figure 5.1: A

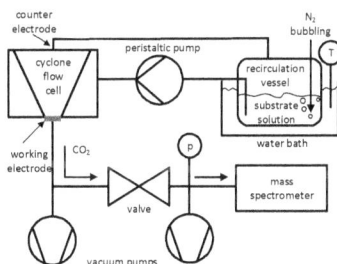

Figure 5.1: Experimental set up including DEMS, cyclone flow cell, and substrate recirculation loop.

peristaltic pump recirculated the substrate solution from the cyclone cell to a glass vessel at $105\,ml\,min^{-1}$. This vessel was constantly purged with high purity nitrogen (99.999% Westfalen AG) in order to remove oxygen and to prevent the accumulation of volatile substances in the bulk solution. The volume of the flow cell is $77\,mL$, the total liquid volume including the recirculation vessel and the tubing is approximately $160\,mL$. FEP (fluorinated ethylene propylene) tubing was chosen as a compromise between chemical inertness and gas permeability.

The temperature in the recirculation vessel was kept at $36\,°C$, resulting in a temperature of $35\,°C$ in the working electrode compartment. The calibration of the MS with CO_2 (99.999 %, Westfalen AG) was done as described elsewhere [116].

A saturated Silver/Silver-Chloride electrode was used as a reference electrode. The counter electrode was made from a platinum wire. It was placed at the outlet of the cyclone cell and separated from the cell volume by a glass frit.

5.2.2 Electrochemical measurements

A Gamry Reference 3000 potentiostat was used for electrochemical measurements. All potentials are reported with respect to the saturated Silver/Silver-Chloride Reference Electrode (Meinsberger Elektroden, Germany, $+0.197\,V$ vs. SHE). Chronoamperometry was recorded at a constant potential of $0.2\,V$, cyclic voltammograms were recorded from $0.2\,V$ to $-0.5\,V$ at the scan rates given below. Two cycles were recorded, they showed only minor deviations. For all CVs, the second cycle is shown. Also potential steps from $0.2\,V$ to $-0.5\,V$ and back to $0.2\,V$ were applied.

5.2.3 Inoculum and media composition

The inoculum was a mixed culture biofilm scraped from an secondary biofilm electrode. No further analysis of the biofilm population was conducted. However, a community analysis was conducted in another recent study [117], where waste water from the same source (Wastewater treatment plant Steinhof, Braunschweig, Germany) was used as initial inoculum, and the same growth conditions and cultivation steps were applied. There the biofilm consisted mainly of *Geobacter anodireducens*. Because of the very similar culturing conditions, it seems likely that *G. anodireducens* also contributed significantly to the biofilm in this study.

Based on previous studies, the reddish colour of the biofilm and the shape of non-turnover CVs (see also section 5.3.2) is attributed to a high share of *Geobacter* species [118, 26, 119].

The substrate solution was prepared with ultrapure water (Millipore Milli-Q 18.2 MΩcm). 1 L of substrate solution contained 10 mmol sodium acetate, 2.69 g NaH_2PO_4, 4.33 g Na_2HPO_4, 0.31 g NH_4Cl, 0.31 g KCl [120], 12.5 ml trace element solution and 12.5 ml vitamin solution [121]. After inoculation, three batch cycles were performed in order to reach stationary biofilm conditions. In the beginning of each batch cycle, the cyclone cell and the recirculation vessel were filled with fresh de-aerated substrate solution. During the batch cycles, no additional substrate was added so that the substrate concentration was continuously decreasing throughout the cycle. The CVs discussed in this work were recorded in the last phase of the third batch cycle. Therefore the results are valid for a stable biofilm that does not change its properties due to biofilm growth. The cell was connected to the vacuum system and the mass spectrometer only in the third cycle.

5.3 Results and discussion

5.3.1 Chronoamperometry

In figure 5.2, the development of current and ion current for CO_2 at m/z = 44 over time are depicted for the last phase of the third batch cycle. The potential of the working electrode is 0.2 V throughout the experiments except for the points marked a to g which will be discussed in the following. Initially the current decreases quickly because the substrate in the reactor is consumed by the anodic reaction which is continuously reducing the substrate concentration. As explained above, new substrate was added at the end of the preceding cycles after the current had dropped; shortly after adding the substrate, the current increased rapidly. Therefore it is assumed that the reaction rate is strongly concentration dependent at the end of the batch cycle when current is low. With decreasing current also the CO_2 signal decreases continuously. The sudden drop in current and ion current after 19 hours at point b was caused by defective electrode contact which was fixed shortly after. From 31 h to 100 h, the current decreases very slowly from 12 µA to 7.5 µA at 100 h with some fluctuations overlaying the downward trend. At all times, the ion current follows the electrical current.

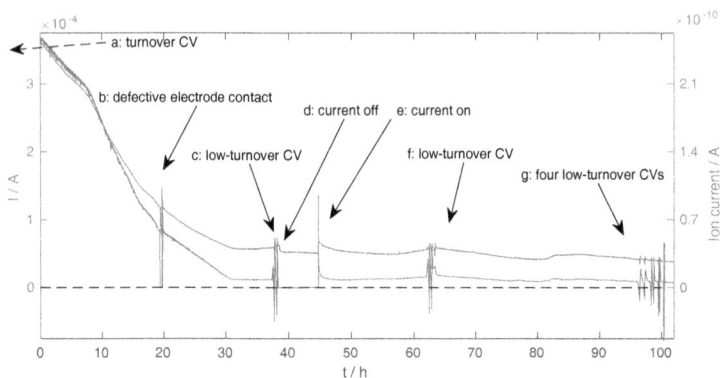

Figure 5.2: Current and ion current at $m/z = 44$ over time. Potential is $0.2\,\mathrm{V}$ except for the points indicated by the arrows. a: turnover CV (recorded 30 h before the data series shown in this figure), b: defective electrode contact (open circuit condition), c: CV one, d: CA interrupted (open circuit condition), e: CA restarted, f: CV two, g: CVs three to six. Ion current is plotted in orange, electrode current in blue.

In figure 5.3, a magnified view of the current and ion current between 85 and 93 hours is shown. It reveals that the sensitivity of the DEMS for CO_2 is remarkably high: at 85.5 hours, for example, the current briefly decreases from 12.3 to $11.8\,\mu\mathrm{A}$ and a change is visible in the ion current signal. The difference of $0.5\,\mu\mathrm{A}$ corresponds to a difference in CO_2 production rate of $6.5 \cdot 10^{-13}\,\mathrm{mol/s}$ assuming complete oxidation of acetate to CO_2. To attain a similar sensitivity by bubbling N_2, for example at $50\,\mathrm{ml/min}$, and analyzing the off-gas, it would be necessary to detect $0.02\,\mathrm{ppm}$ of CO_2 which might be difficult when using for example purity N5 nitrogen that already contains $10\,\mathrm{ppm}$ of unspecified gases. Additionally, stripping CO_2 from a buffered solution is a very slow process and from literature [112] a time constant in the order of $3000\,\mathrm{s}$ can be estimated. As will be shown below, the method presented here allows much faster response times.

When the electrode is situated closely to the porous membrane, which is the case in the set-up, the reaction could be affected by evaporation and subsequent depletion of the reactants [42, 116]. However, in a buffered solution at pH 7, the dissociation of acetic acid (pKs = 4.75) is 99.4 %, resulting in a negligible evaporation rate. Furthermore, the cell was connected to the mass spectrometer and the vacuum system only in the third batch cycle. Current densities recorded under turnover conditions as well as peak positions in the non-turnover CV with

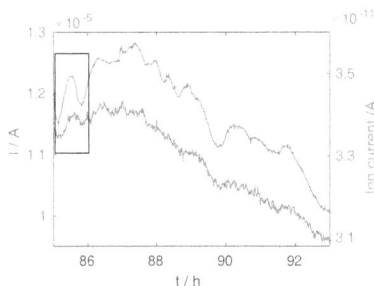

Figure 5.3: Current and MS signal over time from hour 85 to hour 93 at 0.2 V, showing that changes in current of 0.5 µA corresponding to a difference in CO_2 production rate of $6.5 \cdot 10^{-13}$ mol/s cause an observable difference in ion current. Magnified view from figure 5.2. Ion current is plotted in orange, electrode current in blue.

the MS connected, both resemble those without the MS connected very closely. It can be concluded that the measurement set up does not significantly influence the metabolic activity of the biofilm. Consistent with the considerations above, no signal from acetic acid above the background was detected.

Thus the set up is suitable for in-depth analysis of electrochemically active biofilms.

5.3.2 Cyclic voltammetry

In figure 5.4, the MSCV recorded at turnover conditions (labeled 'a' in Figure 5.2) is shown. The current exhibits a typical s-shape with a single point of inflection at -0.337 V and a half saturation potential of -0.320 V. The inflection point is very close to a previously reported value for a pure culture *Geobacter sulfurreducens* biofilm of -0.335 V [26]. The ion current follows the current, but a significant time delay can be observed. From the hysteresis between positive and negative scan of approximately 0.12 V and the scan rate of $1 \, \mathrm{mV \, s^{-1}}$, a delay of 60 s can be calculated. This delay is larger than in DEMS experiments with anorganic catalysts which can be as low as 0.1 s [42]. In the previous study on CO oxidation (see chapter 3) in the same cell it was 1-2 seconds. This increased response time may be attributed to the following differences between the electrodes: The biofilm electrode is thicker than a typical catalyst layer and the transport of CO_2 through the inner and outer cell membranes is slower than desorption of a reaction

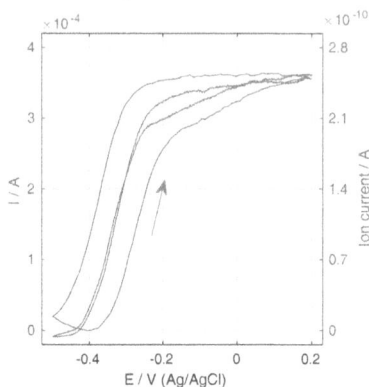

Figure 5.4: MSCV of the biofilm under turnover conditions at 1 mV/s showing a typical s-shape and a high correlation of current and ion current. Ion current is plotted in orange, electrode current in blue.

product from a catalyst surface. All in all, the shape of the MSCV under turnover conditions is as expected.

In total, six MSCVs were recorded under substrate depletion conditions: A first CV was recorded at 1 mV/s at hour 26 (see arrow *c* in figure 5.2) after the current had dropped to 12 µA. After that the cell was in open circuit condition for 6 hours before potential was applied again (*d* to *e*). At hour 62 (*f*), a second CV was recorded at 1 mV/s. Finally, CVs three to six were recorded at hour 96 at scan rates of 0.5, 1, 2, and 5 mV/s (*g*).

In figure 5.5 the third MSCV, which was recorded at a scan rate of 0.5 mV/s, is depicted. MSCVs 4-6, which were recorded at higher scan rates, are given in figures A.6 - A.7 in section A.3.2 in the appendix. In the current signal of the positive scan, a shoulder at -0.415 V, a peak at -0.339 V, a second shoulder at -0.290 V and a second peak at -0.195 V can be distinguished. At higher scan rates, the first shoulder disappears and the second one develops into another peak. From these three peaks of the positive scan and the corresponding peaks and shoulders in the negative scan, three formal potentials can be derived: $E_{f,1}=-0.367$ V, $E_{f,2}=-0.307$ V, $E_{f,3}-0.205$ V. The first two formal potentials are very close to those of the two major redox systems reported in [26] for *Geobacter sulfurreducens* consistent with this species contributing significantly to the biofilm.

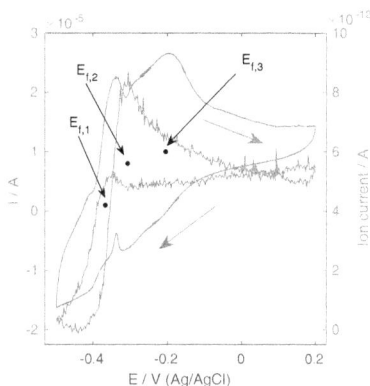

Figure 5.5: MSCV of the biofilm under low-turnover conditions at 0.5 mV/s. Second cycle of the third low-turnover CV from figure 5.2. Ion current is plotted in orange, electrode current in blue.

In all MSCVs under substrate depletion conditions CO_2 production was observed. Therefore they will be referred to as low-turnover CVs, rather than non-turnover CVs in the following. CO_2 production does not follow the current closely in the low-turnover CVs, and a number of noteworthy features can be observed in figure 5.5:

First, in the negative scan, the current becomes negative below -0.2 V whereas the ion current remains nearly unchanged until reaching the first formal potential of -0.367 V. This phenomenon cannot be explained by a time lag in the detection of CO_2 alone: The potential difference between the falling and the rising CO_2 signal around -0.39 V is 35 mV. From this value and the scan rate, a time lag of approximately 35 seconds can be estimated assuming the same delay for the rising and the falling signal. The same delay is obtained from CV four recorded at the double scan rate of 1 mV/s (figure A.5 in the appendix). It corresponds to a potential shift between current and ion current of 17.5 mV at 0.5 mV/s. Thus, CO_2 production must continue at an unchanged rate down to a potential near $E_{f,1}$ even though the total current becomes negative at -0.2 V. Also from the shape of the curve it can be excluded that the negative current results from double layer charging. This means that some components in the ARB must be accepting electrons from the electrode and from acetate metabolism at the same time. This observation is in line with literature where it was shown that the observable high capacitance of electroactive biofilms results from a large number of c-type cytochromes undergoing redox reactions [122].

If the CO_2 production resulted from respiration with trace oxygen in the solution, the ion current would be independent of the potential. Thus the presence of oxygen or other chemical electron acceptors can be excluded as an explanation of this phenomenon. Another reason might be that the cells increase the level of NADH or other reduced components inside the cytoplasm [123]. Continued CO_2 production coupled with NADH accumulation in the cell seems unlikely though, because of the steep decline in CO_2 production below $E_{f,1}$ that appears at various scan rates. If the electrons from acetate oxidation were taken up by a storage mechanism that is not directly coupled to the electrode potential, such as NADH accumulation, it could be expected that the CO_2 production would drop at more cathodic potentials for higher scan rates: the saturation of such a storage is basically a function of time and at higher scan rates a more negative potential would be reached before no more electrons can be accepted. This is not the case. The potential where the CO_2 production begins to drop does not show a clear trend between 0.5 and $5\,mV/s$ and lies between -0.366 and $-0.382\,V$.

A second feature is observed at $-0.35\,V$ in the negative scan: here the ion current rises and reaches a maximum just before dropping. This shows clearly that there are two competing reaction pathways because a single rate limiting electron transfer process cannot account for CO_2 production rising with decreasing potential.

As the CO_2 signal shows a positive peak at $-0.35\,V$, the first oxidation pathways seems to be triggered at low potential, and is not active at high potentials in parallel with the second pathway that is active in the high potential region. The potential range of the peak in CO_2 production suggests that the redox system at $E_{f,1}=-0.367\,V$ is involved in the low-potential pathway. The high-potential pathways lies in the potential range of the redox system with a formal potential of $E_{f,2}$. The existence of two or even more electron transfer pathways has already been suggested in literature [124, 29, 125] for pure cultures of *Geobacter sulfurreducens*. Also, in the potential range of $E_{f,1}$ a low potential electron transfer process has been postulated by [126]. In a recent study employing *in-situ* autofluorescence spectroelectrochemistry [117], it was found that the peak in electrical current in the negative scan, which is also visible in the data at $-0.338\,V$ in figure 5.5, does not only result from a decrease in reduction rate but from an oxidative process. The observed increase in CO_2 production is thus consistent with these previous results. It is nevertheless quite remarkable, considering that the oxidation rate is already strongly limited by substrate availability under low-turnover conditions. However, even at substrate limiting conditions, the substrate concentration does not reach zero. This is evident from the fact that biological reaction rates or growth rates, for example expressed by Michaelis-Menten or Monod-kinetics ($q = q_{max} \cdot c/(c + K_S)$), decrease with decreasing substrate concentration c. The low potential pathway seems to allow the cell to oxidise more substrate at -0.35 V than at higher potentials which usually foster faster oxidation rates. A possible explanation for this might be different half-saturation rate constants K_S of the two pathways. In [127] it was shown that a lower half-saturation rate constant K_S might be associated with a higher energy gain per molecule of substrate. Still, further research efforts are needed to clearly identify the two processes.

Third, the onset potential of CO_2 production is discussed. It can be determined from the positive scan. The ion current begins to rise slowly at -0.42 V. When subtracting a 35 mV shift for the lag of the CO_2 signal, that was discussed above, there is only a difference of 22 mV between the onset potential of CO_2 production and the standard electrode potential of acetate oxidation of -0.477 V [128]. This difference might result from a concentration overpotential: according to the Nernst equation, low acetate concentrations in the biofilm shift the equilibrium potential further upwards (see section A.3.2 in the appendix for calculations). Recently a study conducted by Peng and co-workers [129] also found that an established biofilm still produces current close to the standard potential. The very sharp rise and decline of the CO_2 signal indicates that even though the biofilm consists of a mixed culture, either one species is dominating the biofilm or all species are using a similar electron transfer mechanism.

Fourth, further on in the positive scan, the ion current peaks near -0.3 V at a level that is much higher than the average ion current. This overshooting is remarkable because it indicates a mechanism for the release of substrate from a storage and will be discussed in detail in the next section. In spite of the presence of another current peak, the ion current continuously falls after the peak. The redox system with a formal potential of -0.205 V thus does not seem to be active for the complete oxidation of acetate to CO_2 under these conditions. CVs from a control experiment are shown in section A.3.4 in the appendix.

5.3.3 Quantitative analysis of CO_2 production

A great advantage of mass spectrometry is the fact that it is a quantitative technique. In the following current and ion current, i.e. CO_2 production are quantitatively correlated. Quantifying CO_2 production rates allows to calculate the share of current resulting from substrate oxidation to CO_2 and the share of current resulting from other processes that do not cause CO_2 production such as double layer charging, incomplete substrate oxidation or oxidation of other substances. In figure 5.6, the MS signal is plotted over the current recorded in the first 30 hours of the CA measurement where a rapid depletion of substrate takes place. It can be seen that there is a linear relationship between current and CO_2 production. From the slope of the regression curve, a calibration constant for CO_2 is obtained: the amount of CO_2 produced can be calculated directly from the current and correlated to the ion current since acetate is the only available substrate and thought to be fully oxidised. For the system used here, the calibration constant relating the mass spectrometer signal to the production rate of CO_2 is $K_{cal} = 0.223 \, \mathrm{C \, mol^{-1}}$. A collection efficiency of 41 % was calculated from this value and the MS calibration constant for CO_2 of $0.54 \, \mathrm{C \, mol^{-1}}$ as described in chapter 3.

The intercept of the curve with the y-axis is slightly above zero due to the background of the vacuum system. Changing background currents have already been reported in earlier DEMS studies with a shorter duration [85] and it is common practice to use background-subtracted spectra. In section A.3.1 in the appendix, details on the background subtraction are given.

Since CO_2 is produced in the low-turnover CV, it is to be expected that the time averaged current is positive. If acetate oxidation to CO_2 is the only irreversible reaction taking place, the amount of CO_2 produced calculated by Faraday's law ($\dot{n}_{CO_2} = I \cdot z^{-1} F^{-1}$ with $z = 8$) must be the same as the amount of CO_2 detected by the MS ($\dot{n}_{CO_2} = I_{44} \cdot K_{cal}^{-1}$). In table 5.1, mean values of CO_2 production calculated from the current and from the ion current are summarised for all four scan rates. Please note that no error ranges were calculated and the accuracy of the measurements is less than three digits imply. The mean current is not influenced by the scan rate, and the CO_2 production calculated from the MS signal in all cases matches the one calculated by Faraday's law. This means that acetate oxidation to CO_2 was the only irreversible process in the CV, and reduction and oxidation of cellular redox systems was fully reversible. The agreement of current

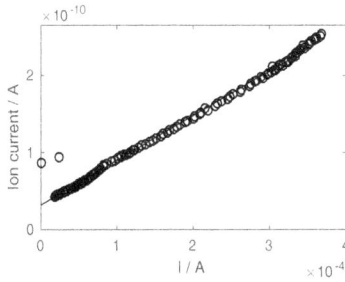

Figure 5.6: Calibration curve recorded during transition from turnover to non-turnover conditions (hours 10 - 30 in fig. 5.2) showing a linear correlation (R^2=0.99) between current and ion current. Data points are plotted as circles. The continuous line is a linear regression function with $R^2 = 0.99$.

and ion current allows to rule out contributions from hydrogen recycling from the counter electrode [130]. Also, no hydrogen was detected by the MS in the experiment.

These results strongly suggest that in any non-turnover CV, where a positive average current density prevails when integrating over the whole CV, some acetate is still left and oxidised during the CV. The average current density can then be used to quantify the rate of irreversible substrate electro-oxidation. Quite remarkably, many of the "non-turnover" CVs reported in literature are actually low-turnover CVs, for instance in Rimboud et al. [33]. These findings suggest that the approximate ratio of reversible and irreversible oxidation currents can be estimated even without a DEMS setup.

5.3.4 Storage mechanisms

In [131] it is reported that substrate is stored by a biofilm in an MFC at high substrate concentrations and oxidised later, under substrate depletion conditions. The study suggests that substrate is stored in shape of storage polymers such as poly-β-hydroxyalkanoates. In this section it will be shown that significant storage also occurs when the electrode potential is too low for immediate substrate oxidation.

To investigate storage capabilities, potential steps were applied and current and ion current were recorded. The potential was stepped from 0.2 V, where electro-

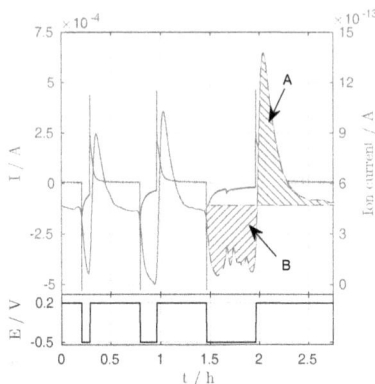

Figure 5.7: Current and ion current upon stepping the potential from 0.2 to -0.5 V for 5, 10 and 30 minutes under low-turnover conditions. Arrows A and B indicate integrals above and below baseline for a step duration of 30 minutes. Ion current is plotted in orange, electrode current in blue.

oxidation of acetate takes place, to -0.5 V, where no acetate electro-oxidation takes place. Then it was held at -0.5 V for a period of 5, 10 and 30 minutes to estimate how much substrate or storage compound was stored over time while no oxidation took place. After the respective holding time, the potential was set to 0.2 V again. In figure 5.7, current and ion current over time are depicted for these potential steps. As expected, the ion current drops rapidly after each step to -0.5 V. After stepping the potential back to 0.2 V, the ion current exhibits peaks that increase in size with the length of the preceding period of low potential. These peaks result from the oxidation of stored substances that were produced during low potential when no oxidation took place. When comparing the integrals of the ion current during the time at -0.5 V and the integral of the peak after switching back to 0.2 V, the storage efficiency can be estimated. Arrows A and B in figure 5.7 show these integrals exemplarily for the last last potential step. In figure 5.8, the integrals of the ion current peaks are plotted over the duration of the preceding period at low potential as red circular markers. The marker indicated by arrow A in figure 5.8 corresponds to the integral area A in figure 5.7. The other points were obtained in the same manner. The peak area increases linearly with the duration of the preceding low potential period, indicating that the storage rate was constant during all three steps. The linear relationship suggests that only a small part of the storage capacity is used even after 30 minutes. Acetate replenishment is not the reason for the observed behaviour: From the recirculation flow rate and the cell volume a hydraulic retention time of 44 seconds can be calculated. Acetate concentration would reach an equilibrium within a time much shorter than the 30 minutes of the last potential step experiment and thus cannot explain the experimental data. The linear relationship in figure 5.7 shows that acetate replenishment does not play a significant role for the shorter steps of 5 and 10 minutes either.

Also in figure 5.8, the integrals of ion current below the baseline during the period at low potential are plotted with blue cross markers. The marker indicated by arrow B corresponds to the integral area B. These integrals match the integrals of the peaks very well. This confirms that all substrate not oxidised at low potential is stored and oxidised at high potential. From the current signal, the storage cannot be deduced directly because of the large capacitive currents that cause a negative current after the steps to -0.5 V. The same measurement was performed under turnover conditions. The respective figures A.12 and A.13 can be found in section A.3.2 in the appendix. Under turnover conditions, storage can be observed also and the amount of storage compound formed still increases linearly with the duration of the low potential period. However, not all substrate that is not oxidised during low potential is converted to a storage compound. The peak area after the positive potential step is only 20% of the area below the baseline during low potential, indicating that the maximum rate of substrate storage is only one-fifth of the maximum rate of substrate electro-oxidation. This rate remains constant for 30 minutes indicating that again only a small part of the storage capacity is utilised.

The potential dependency of the substrate storage can be observed in the MSCV. Actually, two storage mechanisms are visible in the MSCVs which will be discussed further in the following: a) substrate storage and b) charge storage.

In the low-turnover CV in figure 5.5, CO_2 production diminishes at low potentials whereas a large peak can be observed in the positive scan. The average currents and CO_2 production rates over the whole CV, however, approximately equal the values during CA measurements. In table 5.1, the average currents and CO_2 production rates over the CV are reported for all scan rates. Considering that the oxidation rate during CA is limited by substrate availability, the average oxidation rate during the CVs can only be identical to that rate if – just like in

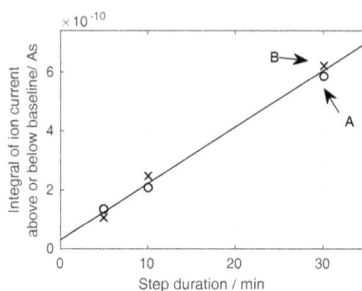

Figure 5.8: Absolute values of the integrals of ion current below the baseline at -0.5,V and integrals of ion current above the baseline directly after stepping back to 0.2 V over the time for which the potential was set to -0.5 V. Arrows A and B correspond to the areas marked in figure 5.7. Integrals above the baseline are plotted as circles, integrals below baseline as crosses. The continuous line is a linear regression function.

the potential step experiments at low substrate availability – all the substrate that is not oxidised in the low potential region is stored as a storage compound and completely oxidised in the high potential region. Table 5.1 also shows that the amount of CO_2 detected by the MS equals the amount of CO_2 calculated from the Faraday current with four electrons transferred per molecule of CO_2 for all CV scan rates. If substances other than acetate or a storage compound synthesised from acetate were oxidised, it is very likely that a different number of electrons per molecule would be transferred. Also it would be difficult to imagine a mechanism that causes the biofilm to metabolise exactly the amount of energy reserves that compensates for reduced acetate oxidation at low potential at four different scan rates.

The storage is visualised in figure 5.9 by plotting current and ion current over time during the MSCV at $0.5 \, \mathrm{mV \, s^{-1}}$. Substrate is stored between 1115 s and 1700 s. Similarly as for the CA measurements, integrals of CO_2 production allow to estimate the amount of acetate equivalent that is stored by integrating the ion current above and below the baseline. In figure 5.9 both areas are indicated by arrows. Because a part of the storage compound is oxidised only in the following cycle of the CV, the area corresponding to oxidation of stored substrate is split into two parts. Using the calibration constant that was defined above, it can be calculated that $4.26 \cdot 10^{-9}$mol of acetate equivalent are stored. At the higher

Table 5.1: Average current, ion current (background corrected), and CO_2 production rates during low-turnover CVs at different scan rates. The values do not deviate significantly from current and ion current during CA measurement right before the CVs

scan rate	mean current	CO_2 from current	mean ion current	CO_2 from ion current
0.5 mV/s	7.78 µA	2.02 $\cdot 10^{-11}$ mol s^{-1}	4.71 pA	2.11 $\cdot 10^{-11}$ mol s^{-1}
1 mV/s	7.77 µA	2.01 $\cdot 10^{-11}$ mol s^{-1}	4.78 pA	2.14 $\cdot 10^{-11}$ mol s^{-1}
2 mV/s	7.77 µA	2.01 $\cdot 10^{-11}$ mol s^{-1}	4.69 pA	2.10 $\cdot 10^{-11}$ mol s^{-1}
5 mV/s	7.40 µA	1.92 $\cdot 10^{-11}$ mol s^{-1}	4.62 pA	2.07 $\cdot 10^{-11}$ mol s^{-1}

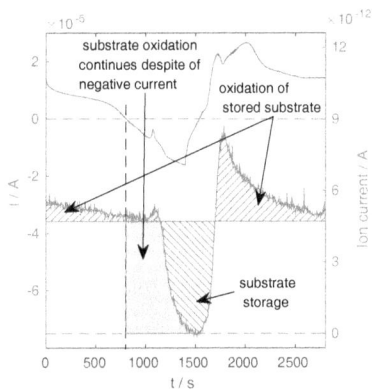

Figure 5.9: Background corrected ion current over time during MSCV at 0.5 mV/s. The average ion current is approximately the same as the ion current during CA right before the CV. The hatched areas below and above the average line have the same size and are proportional to amount of substrate converted to a storage compound at low potential and oxidised later on at higher potential. Ion current is potted in orange, electrode current in blue.

scan rates, the amount of storage is lower because the period of time where the potential is too low for substrate oxidation is shorter. The formation of the storage substance does not go along with CO_2 production. At high potentials it is completely oxidised to CO_2. In [108] it is hypothesised that ARB could continue substrate breakdown under production of NADH in absence of an extracelluar electron acceptor. This step, however, goes hand in hand with CO_2 production. As no CO_2 production is observed at very low electrode potentials there must exist another mechanism that takes up acetate without liberation of CO_2. This is consistent with results from Song et al. [132] who reported that the citric acid circle in *Geobacter sulfurreducens* is regulated down at lower current densities. The accumulation of storage polymers as suggested in [131] seems more likely considering the data because, for instance, the transformation of acetate via Acetyl-CoA and Acetoacetyl-CoA to PHB would not release CO_2. Elucidating the exact nature of the storage mechanism is out of scope of this work as it needs further refined analysis.

Charge storage occurs after 800 s when the current begins to flow in cathodic direction. The grey area under the ion current curve represents the amount of CO_2 that is produced in spite of a cathodic current. During this time, electrons from acetate oxidation and from the cathodic electrode current must be stored in the biofilm. From the integral of current (5.86 mC), which corresponds to stored e^-, and of CO_2 production (3.81 mC), which would release e^- as well, a total charge storage capacity of 9.67 mC can be calculated.

5.4 Conclusions

In this chapter, it was shown that DEMS constitutes a valuable technique for analysing biofilms of electrode respiring bacteria in-vivo and online. It was demonstrated how to gain insights into the coupling of extracellular electron transfer and metabolic activity: MSCVs were recorded under turnover and under substrate depletion conditions. The onset of complete oxidation of acetate to CO_2 is a few mV above the equilibrium potential. Two competing extracellular electron transfer mechanisms have been directly identified in low-turnover CVs, the first one is active at potentials greater than 0.35 V, the second one below that potential. Furthermore, the existence of two significant storage mechanisms has been shown: 1) A charge storage mechanism that allows substrate oxidation to proceed at a constant rate in spite of current flowing in cathodic direction during the negative scan of the low-turnover CV. 2) A carbon storage mechanism that allows the biofilm to take up acetate at an unchanged rate even though the oxidation to CO_2 stops at very low potentials. Whereas this work lays a basis for quantitative analysis of substrate storage mechanisms, many more questions are open concerning the factors that determine these processes. It was observed that the storage process does not cause CO_2 production. When the potential is too low to allow oxidation of acetate, at low substrate availability, all substrate that is not oxidised is stored in the biofilm. When the potential is too low to allow oxidation of acetate, at high substrate availability, the storage rate was observed to be 20 % of the substrate oxidation rate at high potential.

These results suggest that, under certain dynamic conditions, charge transfer is only loosely connected to acetate consumption and CO_2 production. This is an important fact to be further explored and to be considered when evaluating dynamic electrochemical measurements such as CV or EIS. A better understanding of factors that influence storage capacities might also enable a more flexible operation of bioelectrochemical cells to meet requirements of fluctuating energy demands.

Although a number of new insights into biofilm electrodes could be gained, no modelling of the biofilm DEMS cell is conducted. The results presented above show that the biofilm behaviour is largely determined by biological processes and substrate storage processes that are not easy to describe by common modelling approaches for electrochemical systems. Additionally, it is not clear if the processes under turnover conditions and non-turnover conditions are identical, and batch operation further complicates modelling studies because the biofilm is continuously changing with time. For these reasons no attempt at describing the DEMS experiments on an acetate oxidising biofilm with a physical model is made. In the next chapter, a BES that is more suitable for quantitative description by a mathematical macro kinetic model will be analysed.

Chapter 6

Glycerol oxidation in a bioelectrochemical system[1]

In this chapter it will be demonstrated how reaction pathways and kinetics of the comparatively complex molecule glycerol can be elucidated by dynamic methods and additional concentration measurements in a continuously operated reactor. Glycerol oxidation in a BES is a challenging reaction because for mixed bacterial cultures, pathways and products are not available in literature. Learning from the results of the previous chapter, the focus is placed on species conversion and current generation instead of the electron transfer mechanism. Since the time scale of these processes is longer than that of electron transfer, HPLC will be used as an analytical technique in this chapter offering the advantages of higher sensitivity and a larger range of species that can be detected.

6.1 Introduction

Understanding reaction pathways and process interactions within BES running on complex substrates is an important step towards performance optimization. Many reports on complex substrates which are degraded by a defined artificial community of two or more species were able to assign clear roles to single bacterial species [133, 134, 135]. However, the performance of these cultures is often low compared to mixed cultures, making the latter more attractive for applications.

Even though a wide range of techniques is available for analysis of biofilms in BES [107], often little is known about metabolic pathways for complex substrates.

---

[1]Parts of this chapter have been published in F. Kubannek, C. Moß, K. Huber, J. Overmann, U. Schröder, U. Krewer, Concentration Pulse Method for the Investigation of Transformation Pathways in a Glycerol-Fed Bioelectrochemical System, Frontiers in Energy Research 6 (2018) and in F. Kubannek, U. Krewer, Modelling and parameter identification for a biofilm in a microbial fuel cell, Chemie Ingenieur Technik, accepted manuscript (2019).

There is a knowledge gap concerning pathways and kinetics for degeneration of complex substrates in BES by mixed cultures. Quantitative information is especially rare and there are also no standard methods available to close this gap because the types of interaction strongly depend on the individual BES.

The goal of this chapter is to demonstrate a methodology to extract qualitative information on reaction pathways and process interactions as well as quantitative information on rate constants by evaluating the dynamic response of a BES.

In the field of anaerobic digestion, concentration pulse responses have been evaluated quantitatively using concentration data over time: [136] successfully parametrised a model for anaerobic digestion of olive pulp with data obtained from pulse injection of various intermediates. Based on this promising approach, the concentration pulse method is further developed to analyse BES and extract quantitative information.

In BES literature, only responses of the current density to concentration pulses have been discussed in-depth, whereas concentration data have not been analysed in detail. [137] used acetate pulses to show that acetate availability was limiting current production in an MEC treating landfill leachate. Similarly, [138] demonstrated with pulse experiments that ethanol was not consumed directly in an MEC, but fermented to acetate first.

Glycerol is selected as a promising example substrate to apply the concentration pulse methodology for a number of reasons: In recent years, glycerol utilization technologies have received increasing attention because glycerol is the major by-product from biodiesel production [139, 140]. About 10 % (w/w) glycerol is produced for every unit of biodiesel [141]. Glycerol is also an interesting substrate for BES, and several approaches have been reported in literature, including electricity generation [142, 143, 144, 145], 1,3-propanediol production [146, 147], and hydrogen production [110, 148, 149, 109, 150, 151].

A number of fermentation products have been identified in BES utilizing glycerol as carbon source, e.g. lactate, butyrate, ethanol, methane, acetate, propionate, 1,3-propanediol. A more detailed overview is given in table 6.1.

Still, while metabolic pathways for fermentation of glycerol have been thoroughly discussed in literature [154, 140, 155, 146], it is not easy to determine which pathways are active in a given glycerol-fed BES nor which of the fermentation products contribute to electricity generation and which do not. Thus glycerol is a promising example substrate to demonstrate the methodology.

Table 6.1: Intermediates detected in glycerol-fed BES in literature. Note that an empty field includes the cases "not found" as well as "not determined" and "not reported". Reprinted from [152], SI (CC BY 4.0).

	COD	glycerol	valerate	butyrate	1,3-PDO	lactate	propionate	ethanol	acetate	formate	CO_2	CH_4	H_2	culture	type	mode	others
[109]		x			x		x	x	x		x	x	x	mixed	MEC	B	
[110]	x	x					x	x	x		x	x	x	mixed	MEC/MFC	FB	
[147]		x	x	x	x		x	x	x	x	x	x	x	mixed	EF	C	*
[148]		x				x		x	x	x	x		x	Enterobacter aerogenes	MEC	B	
[153]		x		x	x	x	x	x	x	x				mixed	EF	B	**
[143]	x	x	x	x			x		x				x	mixed	MFC/MEC	B	***
[150]		x	x	x				x	x					mixed	MFC	B	
[151]		x	x	x	x		x	x	x		x		x	mixed	MEC	C	
[135]		x												Shewanella Oneidensis & Klebsiella pneumonae	MFC	B	
[144]	x													mixed	MFC	B	
[145]	x													mixed	MFC	B	
[149]												x		mixed	MEC	B	
[142]														Bacillus subtilis	MFC	B	
This work		x	x	x	x	x	x	x	x	x	(x)	(x)		mixed	MFC	C	

* = hexanoic acid, ** = succinate, *** = isocaproic acid, caproic acid, octanoic acid and isovalerate, EF = electro-fermentation, MFC = microbial fuel cell, MEC = microbial electrolysis cell, 1,3-PDO = 1,3-propanediol, B = batch, C = continuous, FB = fed-batch

Here a continuously operated glycerol-fed BES is set up and concentration pulse responses are recorded by observing the transients of current and concentrations after the injection of intermediates at high concentrations. With this methodology, the pathway from glycerol to current are identified and conversion efficiencies for each intermediate are calculated. It is shown how these experiments can be used to extract rate constants for substrate conversion and the dynamic system behaviour is characterised. Finally, community analysis links the pathways to key players in the microbial community.

6.2 Material and methods

6.2.1 Reactor setup and operation

In figure 6.1, the experimental setup is depicted. The experiments were conducted in a heating cabinet (L) at a constant temperature of 35 °C. The reactor consisted of a 5-necked round bottom glass flask (A) that was agitated by a magnetic stirrer (I). The working electrode (B) was made from a graphite rod (CP Handels-GmbH, A=6 cm², d = 1 cm). The counter electrode (D) was made from the same material as the anode. The reactor was operated in half-cell mode under potentiostatic control with a ZiveZ5 potentiostat (WonATech, South Korea). All potentials are reported with respect to a saturated Ag/AgCl reference electrode (C, Meinsberger Elektroden, Germany, +0.197 V vs. SHE). Chronoamperometry was recorded at a constant potential of 0.2 V, cyclic voltammograms were recorded from 0.2 V to -0.5 V at 1 mV/s. Before starting an experiment, the reactor was filled with 180 ml of a deaerated solution consisting of a glycerol solution and a vitamin and mineral solution. Details of the media composition are provided in the next section. After inoculation, the reactor was kept in batch mode until current production was detected. A three-channel peristaltic pump (K) was used to continuously feed the reactor with glycerol solution (E) and vitamin and mineral solution (F). Nitrogen-filled gas bags (M) were attached to the bottles to avoid under-pressure in the feed bottles that could cause air to enter. The ratio of the volumetric flow rate of glycerol solution to vitamin and mineral solution was 1:0.85 at all times. All glycerol inlet concentrations and flow rates are reported with respect to the total volume flow after mixing both solutions in the following.

The liquid level was controlled via a balance (J), whose signal was used to control the third channel of the peristaltic pump, that was pumping solution out of

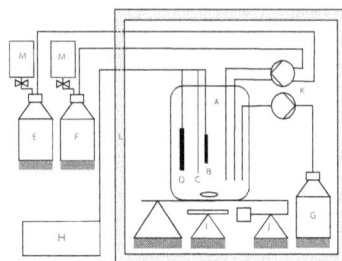

Figure 6.1: Experimental setup. A: reaction vessel, B: working electrode, C: reference electrode, D: counter electrode, E: glycerol solution bottle, F: vitamin and trace element solution bottle, G: effluent bottle, H: potentiostat, I: magnetic stirrer, J: balance, K: three-channel peristaltic pump, L: heating cabinet, M: nitrogen gas bags. Reprinted from [152] (CC BY 4.0).

the reactor into the effluent bottle (C). The hydraulic residence time was 28.8 h for all experiments, apart from pretests, for which hydraulic residence time was varied between 5.4 and 21.6 h. HPLC samples were taken at the end of the outlet tube and filtered through a 0.2 μm syringe filter before HPLC analysis. From the volume of the tube and the flow rates, a mean residence time of 15 minutes was calculated for the outlet tube and taken into account during sampling. The reactor was inoculated in batch-mode. After rising current production indicated a successful colonization, continuous operation was started. The microbial community that developed in the anodic biofilm and in the planktonic phase was determined by RNA sequencing after two months of operation (see section 6.2.4). For the concentration pulses, approximately 2 ml concentrated solution was injected directly into the 180 ml solution within the reactor through a silicone rubber stopper. The liquid level was not influenced significantly by the injection and the parameters of the pump controller were chosen in a way that the 2 ml injection did not cause a rapid change in outlet flow rate. A first set of concentration pulse experiments was performed under closed-circuit potentiostatic operating conditions beginning 29 days after inoculation, and a second set of pulse experiments was performed under open circuit conditions beginning 114 days after inoculation.

6.2.2 Inoculum and media composition

Buffered glycerol solution was produced by dissolving glycerol (PanReac Appli-Chem) in a pH 7 buffer solution containing 2.69 g NaH_2PO_4, 4.33 g Na_2HPO_4, 0.31 g NH_4Cl, and 0.31 g KCl per litre [120]. The glycerol solution was boiled prior to the experiments for deaeration and disinfection. Vitamin and trace element solution was made by mixing 12.5 ml trace element mixture and 12.5 ml vitamin mixture with 1025 ml of deionised water [121]. The vitamin and trace element solution was deaerated with high purity nitrogen for at least 20 minutes prior to the experiments. For the concentration pulse experiments, 50 ml of concentrated solutions were produced by dissolving sodium propionate (Sigma-Aldrich), sodium acetate (Merck), 1,3-propanediol (Sigma-Aldrich), or glycerol (PanReac AppliChem) in buffer solution. The substances were chosen after a pretest (see section SI1 in the supporting information), and the solutions were bubbled with N_2 for at least 20 minutes prior to injection.

The inoculum used in this study was taken from a mixed-culture biofilm dominated by Geobacter anodireducens/sulfurreducens which was enriched from waste water (from the Wastewater treatment plant Steinhof, Braunschweig, Germany). Details on the enrichment procedure have been described by [156]. Glycerol was the sole initial carbon source in the reactor. No community analysis of this initial inoculum was performed.

6.2.3 Concentration measurements

Quantitative concentration analyses were performed by HPLC, with a refractive index (RI) detector (Spectrasystem P4000, Finnigan Surveyor RI Plus Detector, Fischer Scientific), a HyperREZ XP carbohydrate H+8 µm (S/N: 026/H/012-227) column, and H_2SO_4 (0.01 mol L^{-1}, flow rate 0.5 mL min^{-1}) as eluent. The column and the RI detector were operated at 25 °C. Samples were collected from the reactor effluent, and bacteria were removed by filtering with a 0.2 µm pore size filter. No other pretreatment was performed. Some HPLC samples were measured repeatedly and the concentration values showed only minor deviations which were no larger than 0.02 mmol L^{-1}.

6.2.4 Bacterial community analysis[2]

After 67 days of operation part of the biofilm was scraped from the anode with a sterile doctor's blade for sequencing analysis of the anode biofilm community. At the same time, a sample from the solution inside the reactor was taken to analyse the microbes in the planktonic phase. After taking the samples, the reactor was sealed again and operation was continued.

A modified protocol of Lueders and Friedrich [157] was employed to extract RNA from the biofilm and the planktonic phase samples. The precipitation step of the nucleic acids in polyethylene glycol was prolonged to 90 min instead of 30 min and the nucleic acids were finally resuspended in 50 µL EB buffer instead of 100 µL EB buffer as originally described in [157]. The DNA was removed by the application of the Qiagen RNeasy MinElute CleanUp Kit and of the Thermo Fisher RNase-free DNase I digestion Kit according the manufacturers' instructions. Subsequently, a potential contamination with remaining DNA was assessed with the Qubit^{TM} dsDNA HS Assay Kit and the concentration of RNA was checked with the Quant-iT^{TM} RiboGreen^{TM} RNA Assay Kit. After the reverse transcription of the RNA into cDNA, amplicons of the V3 region of the 16S rRNA gene were prepared with the primer pair 341f and 515r as described by [158]. The quality of the Bartram libraries was checked with a 2100 Bioanalyzer (Agilent Technologies, Santa Clara, CA, U.S.) and the subsequent sequencing was performed in 100 bp paired end mode on a HiSeq2500 (Illumina, San Diego, CA, U.S.). The generated sequences were processed with an amplicon analysis pipeline after the quality of the raw reads had been checked by FastQC version 0.10.1 (Simon Andrews; www.bioinformatics.babraham.ac.uk/projects/fastqc/). After the trimming of the forward and the reverse reads to a length of 100 bp, the raw sequence data were purified from potential primer dimers by a JAVA program called DimerFilter. Fastq-join [159] joined the forward and reverse reads with a 20 percent mismatch and a minimum overlap of 6 bp. FASTA converted sequence files were subsequently checked with Uchime (Usearch 5.2.32 [160]) against the gold database provided by ChimeraSlayer (drive5.com/otupipe/gold.tz) and the RDP classifier 2.10.1 [161, 162]. The RDP classifier 2.10.1 [161, 162] with a confidence value of 0.5 which is recommended for short read amplicon data was employed to perform taxonomic dependent analyses of the bacterial community.

---

[2]The community analysis was performed by Katharina Huber from the Leibniz Institute DSMZ - German Collection of Microorganisms an Cell Cultures.

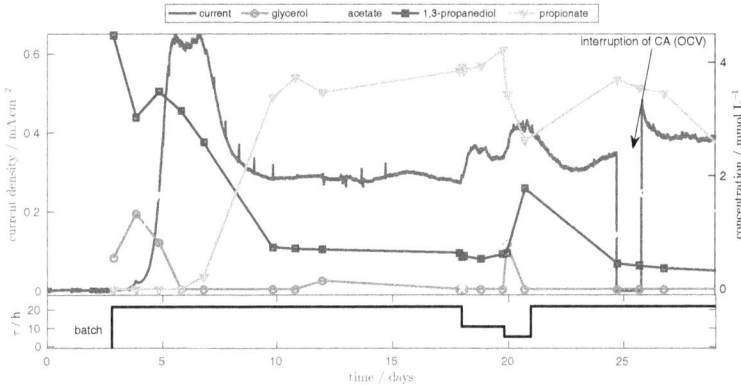

Figure 6.2: Current and concentrations over time during an experiment with a glycerol inlet concentration of 7 mM. Residence time was stepped from 21.6 to 10.8 and 5.4 h by adjusting the flow rate. Reprinted from [152], SI (CC BY 4.0).

6.3 Results and discussion

6.3.1 Pretest: Step changes in residence time

The substances for concentration pulse experiments were selected based on previous experimental runs. Intermediates that were detected in the previous experiment were injected for the concentration pulses. The development of current and concentrations over time for a previous experiment with an inlet concentration of $7.55 \, \mathrm{mmol \, l^{-1}}$ is depicted in figure 6.2. After a lag phase, the current increases rapidly, reaches a maximum of $0.65 \, \mathrm{mA \, cm^{-2}}$, and falls back to a steady state value of $0.3 \, \mathrm{mA \, cm^{-2}}$ on day 13. The peak current density is comparable to pure *Geobacter* spp. biofilms metabolizing acetate [163]. Propionate and 1,3-propanediol are the most prominent intermediates in this steady state. From day 21 onwards, the residence time was reduced stepwise from $21.8 \, \mathrm{h}$ down to $10.8 \, \mathrm{h}$ and $5.4 \, \mathrm{h}$. While the current slightly increased upon each reduction of the residence time, the concentrations of intermediates only changed when the residence time was reduced to $5.4 \, \mathrm{h}$. On day 27, the current was stepped down to zero (open circuit conditions) for one day. After these changes, the system was reverted to the original operational conditions. All in all, the results of this experimental series do not shed much light on the mechanisms of glycerol electro-oxidation. They did help, however, to identify intermediates for the pulse experiments and

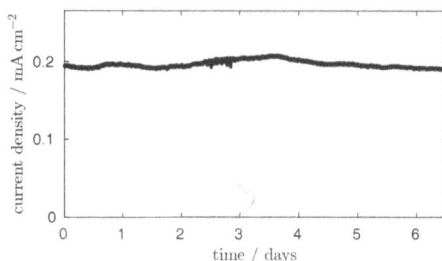

Figure 6.3: Current over time after chemostatic conditions are established. Residence time is 28.8 h, glycerol inlet concentration 1.8 mM, anode potential 0.2 V. Reprinted from [152] (CC BY 4.0).

to choose residence time and inlet concentration for the experiments reported in following sections.

6.3.2 Concentration pulse responses under closed-circuit potentiostatic conditions

In this section, the results of the concentration pulse experiments under electrical closed-circuit conditions and continuous flow are discussed. The reaction pathway that is suggested based on the experimental results is summarised in section 6.3.4.

In figure 6.3, the typical fluctuation of the current density in the continuously operated reactor over time is shown. After chemostatic conditions are established, a steady-state current density of approximately $0.2 \, \text{mA cm}^{-2}$ is reached. The fact that a steady state can be maintained with only minor fluctuations and the fact that the system returns to steady state after each concentration pulse are important conditions for evaluating the concentration pulse experiments which are discussed in this section.

In figure 6.4 (A), the system's response to a pulse in acetate concentration is depicted. Current density is plotted on the left axis, concentration values on the right axis. After injecting acetate, the current density rises rapidly, within 4 minutes, from $0.185 \, \text{mA cm}^{-2}$ to $0.235 \, \text{mA cm}^{-2}$. Subsequently, the current density continues to increase linearly for one day to a level of $0.72 \, \text{mA cm}^{-2}$, which is sustained for 7.2 h. After this plateau, the current density drops because of the decreasing acetate concentration. The acetate concentration is falling because acetate is washed out of the reactor, which has a residence time of 28.8 h, and is consumed by the microorganisms. From the almost instantaneous increase in current density and the absence of any other concentration peaks it is concluded that acetate is directly oxidised in the biofilm without other intermediates being involved. The maximum current density is close to the observed current density in the same reactor using acetate as the only carbon source (data not shown). Also, the community analysis (see section 6.3.8), which shows a high share of *Geobacter* sp. in the biofilm, supports the conclusion that acetate is oxidised directly in the biofilm. The comparatively slow linear increase from $0.235\,\mathrm{mA\,cm^{-2}}$ to $0.72\,\mathrm{mA\,cm^{-2}}$ is attributed to biofilm growth because of the duration of about one day. Biofilm growth is analysed quantitatively below. After each pulse the current density returns to its steady state value within three to five days. Thus biofilm growth is a reversible process and the biofilm returns to its original performance level after the injected substances have been washed out of the reactor.

Next, the concentration pulse response for 1,3-propanediol was recorded (figure 6.4 (B)). The current density rises quickly after a short lag time and reaches $0.394\,\mathrm{mA\,cm^{-2}}$ at 21 hours. After a short drop to $0.370\,\mathrm{mA\,cm^{-2}}$, a second peak is reached 13.75 hours after the first one. Subsequently, the current density approaches its original value again. The peaks in current density correspond to a period of high acetate concentration. From this it can be concluded that current production from 1,3-propanediol proceeds mainly via acetate. After the 1,3-propanediol pulse, glycerol concentration rises first, peaking after approximately 8 hours. Acetate and propionate concentrations follow with a slight delay and reach their maxima approximately 14 hours after the concentration pulse, which is consistent with the time of the maximum current. Thus, either 1,3-propanediol is converted to glycerol, which is then converted to acetate, or 1,3-propanediol inhibits glycerol metabolism, causing the concentration to rise because of the steady inflow of glycerol.

Similar effects can be seen after the glycerol concentration pulse in figure 6.4 (C). After the pulse, concentrations of 1,3-propanediol, acetate and propionate all rise with a very small time delay. While acetate and propionate can be detected over a period of 36 hours, the concentration of 1,3-propanediol diminishes more quickly. Glycerol concentration falls rapidly and reaches a local minimum of $0.26\,\mathrm{mM\,L^{-1}}$ 5 hours after the pulse. After this minimum, glycerol concentration increases again up to $0.64\,\mathrm{mM\,L^{-1}}$. This concentration increase goes along with the drop

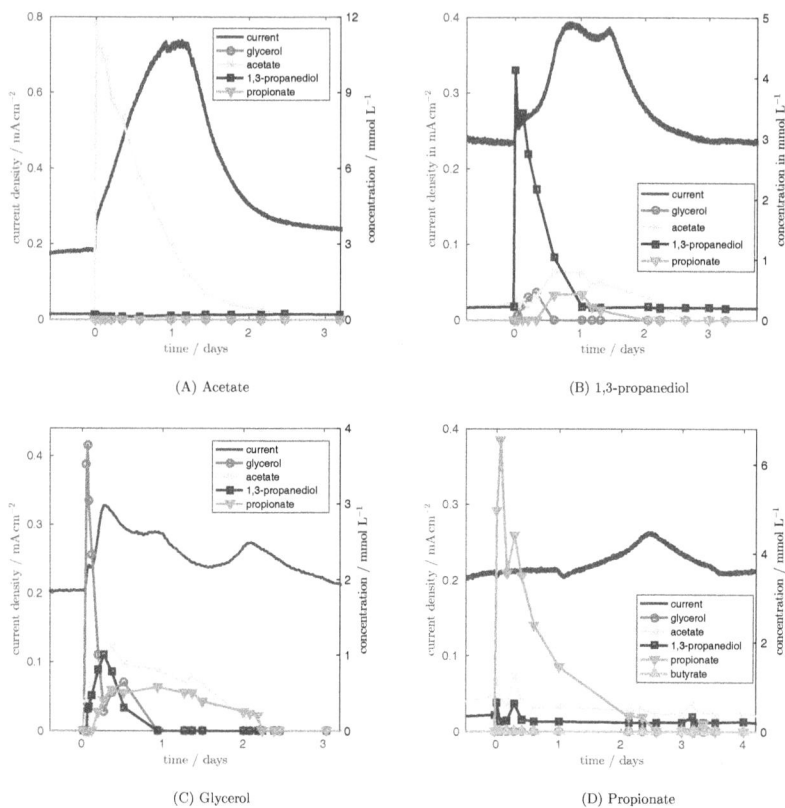

(A) Acetate

(B) 1,3-propanediol

(C) Glycerol

(D) Propionate

Figure 6.4: Concentration and current responses to an acetate (a), a 1,3-propanediol (b), a glycerol (c), and a proprionate (d) concentration pulse at t=0. Residence time is 28.8 h, glycerol inlet concentration 1.8 mM, anode potential 0.2 V. Reprinted from [152] (CC BY 4.0).

in 1,3-propanediol concentration, a similar interaction as was observed for the 1,3-propanediol pulse. At the same time, acetate and propionate concentrations do not fall significantly, rendering a reaction from 1,3-propanediol to glycerol as the more likely cause, compared to inhibition of glycerol degradation by 1,3-propanediol. This reaction has not been in the focus of previous work because glycerol fermentation to 1,3-propanediol is viable from an economic point of view,

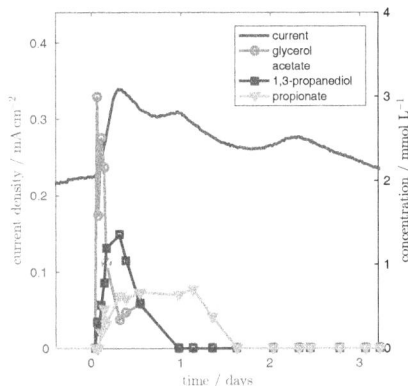

Figure 6.5: Concentration response to a glycerol concentration pulse at t=0. Residence time is 28.8 h, glycerol inlet concentration 1.8 mM. Reprinted from [152] (CC BY 4.0).

whereas the reverse reaction is not. The lack on literature data for this reaction necessitates further research to explain the observed conversion in detail. The glycerol concentration pulse was repeated to check whether the reactor history has an influence on the concentration pulse responses and thus whether the microbial consortium changed in its functionality. The repetition experiment shown in figure 6.5 indicates that this is not the case.

5.3 hours after the glycerol concentration pulse, the current density increases to $0.328\,\mathrm{mA\,cm^{-2}}$. This maximum coincides with the maximum acetate concentration. A second, lower peak appears after 21.1 hours, while acetate concentration is still high. 49 hours after the pulse, a third peak with a height of $0.234\,\mathrm{mA\,cm^{-2}}$ appears, even though metabolite levels in the effluent are not elevated any more. This suggests that the charge released during the third current peak is stored in some intra-cellular intermediate that is not detected by HPLC. Such dynamics in the current production of electroactive biofilms resulting from the storage of intra-cellular intermediates have been reported previously [164].

The response to the propionate concentration pulse, which can be seen in figure 6.4 (D), suggests that propionate might be involved in the pathway that causes the delayed increase in current. After injecting propionate, the current density remains unchanged for about 30 hours. After that, the current density slowly rises to $0.265 \, \text{mA} \, \text{cm}^{-2}$ at hour 60 after the pulse. At the time of the maximum current density, no elevated intermediate concentration can be detected. Since the time delay and the shape of the peak are similar to the third peak from the glycerol concentration pulse, it seems likely that a similar mechanism causes this behaviour. It is assumed that an intracellular metabolite is formed from propionate, which is either converted to current via acetate that is produced from propionate in a slow reaction, or oxidised directly after a slow adaptation of the biofilm. Both mechanisms have been described in literature: [165] reported good performance of a propionate-fed MFC in abscence of *Geobacter* spp. showing that a direct electrooxidation of proprionate is possible. On the other hand, [166] suggested that current from propionate would be generated via acetate.

6.3.3 Concentration pulse responses under open-circuit conditions

The results discussed so far indicate that there is a clear division of labour between the planktonic phase and the biofilm. Fermentation of glycerol to acetate takes place in the planktonic phase and acetate is oxidised by anode respiring bacteria in the biofilm. To check how the fermentative pathway is influenced by the anode respiring bacteria, pulse experiments were repeated under open-circuit conditions. The flow rate and glycerol inlet concentration were at the same values as before. The pulse responses to 1,3-propanediol, glycerol and acetate under open-circuit conditions are shown in figure 6.6.

Overall, the concentration transients match those under closed-circuit conditions well, with the main exception of acetate. Acetate concentration rises by approximately $0.95 \, \text{mmol} \, \text{L}^{-1}$ after switching to open circuit conditions.

By Faraday's law it can be calculated that the concentration increase directly after switching to open-circuit conditions matches the current density under closed-circuit conditions:

$$I = z \cdot \text{F} \cdot \dot{n}$$
$$= 8 \cdot 96\,485 \, \text{C/mol} \cdot 0.104 \, \text{mL/min} \cdot 0.95 \, \text{mmol/L}$$
$$= 1.27 \, \text{mA} \, \widehat{=} \, 0.21 \, \text{mA} \, \text{cm}^{-2}$$

This confirms the findings from the concentration pulse experiments under closed-circuit conditions: the oxidation of acetate to CO_2 is the main electrochemical

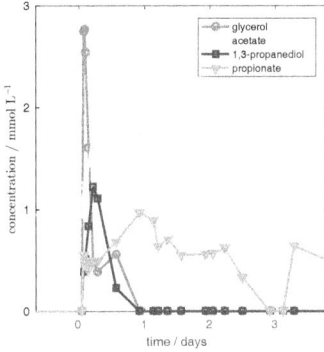

a) Concentration response to a glycerol concentration pulse.

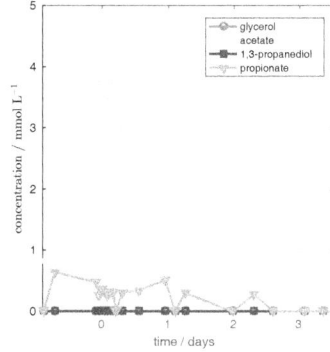

b) Concentration response to an acetate concentration pulse. Glycerol and 1,3-propanediol were not detected. Propionate probably results from the preceding glycerol pulse.

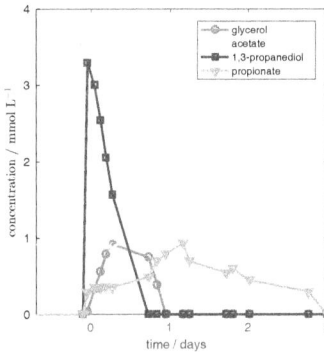

c) Concentration response to a 1,3-propanediol concentration pulse.

Figure 6.6: Concentration responses to concentration pulse at t=0 under open circuit conditions. Residence time is 28.8 h, glycerol inlet concentration is 1.8 mM. Reprinted from [152], SI (CC BY 4.0).

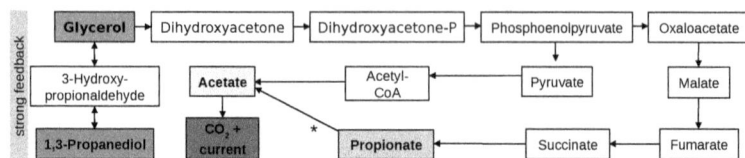

Figure 6.7: Main active pathways and interactions during glycerol electro-oxidation in an MFC. Substances in coloured boxes were measured directly, intermediates are based on [140]. The pathway from propionate to acetate, denoted by a * is unclear. Reprinted from [152] (CC BY 4.0).

reaction and requires the anode as a terminal electron acceptor. All other reactions are chemical reactions and are not strongly influenced by the potential at the given conditions. The community analysis (see section 6.3.8) also supports this interpretation. At higher organic loading rates, where accumulation of fermentation products inhibits a further substrate breakdown, different behaviour may occur.

The acetate concentration is only $0.3 \, \mathrm{mmol \, L^{-1}}$ after 30 days under open-circuit operation which probably results from adaptation of the bacterial community to the new conditions. A possible explanation would be the slow proliferation of methanogens which are outcompeted by the anode respiring bacteria under closed-circuit conditions.

The quantitative match between the drop in current density and the increase in acetate concentration also indicates that hydrogen recycling from the cathode to the anode does not contribute significantly to the observed current density, even though no special precautions were taken to prevent this effect.

6.3.4 Pathways active in the bioelectrical system

In the previous sections, it was shown that current production mainly results from acetate electrooxidation. Since all pulsed substances resulted in a response of the current density, it is assumed that they are all partly converted to acetate via biochemical reactions. Furthermore, glycerol and 1,3-propanediol are directly linked. Thus, it can be concluded from the concentration pulse experiments and from literature studies that the pathways shown in figure 6.7 dominate the processes in the BES. The substances in coloured boxes were measured directly, the others are based on literature [140, 154]. Conversion of glycerol to 1,3-

propanediol via 3-hydroxy-propionaldehyde does not result in current production directly, but a backward reaction to glycerol is triggered at higher 1,3-propanediol concentrations. Current production most likely proceeds via phosphoenolpyruvate, acetyl-coenzyme A (acetyl-CoA) and acetate. Acetate electrooxidation is the only part of the reaction scheme taking place in the biofilm, all other reactions are biochemical reactions taking place in the planktonic phase. Propionate is most likely produced in a pathway starting from phosphoenolpyruvate via succinate. The observed pathway from propionate to current that leads to a long delay of the current response in the propionate pulse remains unclear.

The presented pathway is valid for the given experimental conditions. It is intended as an overview on the main active pathways in the glycerol BES at the given operating conditions and demonstrates the principle capability of the concentration pulse method. To draw general conclusions on pathways in certain bacterial species, defined cultures need to be used.

However, the concentration pulse responses allow to identify important interactions and dependencies in a running reactor without performing multiple batch experiments. In combination with literature, the analysis of the dynamic system response to concentration pulses allowed us to identify probable pathways.

6.3.5 Conversion efficiencies

Apart from qualitative information, quantitative information can also be extracted from the pulse experiments. In table 6.2, the values of the coulomb efficiency (CE), calculated according to equation 6.1, are reported for each pulse experiment. Similarly to previous reports [150], the injected amount is corrected for the charge flushed out with the effluent for CE calculation because the continuous flow leads to a large share of the injected substrate i being flushed out before it can be oxidised:

$$CE_i = \frac{\int_0^{t_{end}} (I - I_0)\,dt}{n^{in} z_i \cdot F - \sum_j \int_0^{t_{end}} \dot{V}(c_j^{out} - c_{0,j}^{out}) z_j F\,dt} \tag{6.1}$$

I is the current during a pulse experiment, I_0 is the steady state current. Integration is performed in the time interval $[0 - t_{end}]$ that ranges from the time of the pulse until concentrations and currents have returned to steady state. The integral in the denominator is thus the additional charge resulting from the concentration pulse which is transferred to the electrode above the steady state level. n_i^{in} is

Table 6.2: Coulomb efficiencies for all concentration pulse responses and steady state glycerol operation. Reprinted from [152] (CC BY 4.0).

injected substrate	charge injected	charge from current	charge flushed out	coulomb efficiency
acetate	1,635 C	434 C	982 C	66.5%
1,3-propanediol	1,150 C	112 C	585 C	19.9%
propionate	1,595 C	30 C	1,357 C	12.5%
glycerol	919 C	95 C	564 C	26.7%
steady state (glycerol)	383 C/day	104 C/day	0.18 C/day	28.4%

the amount of substrate i injected during the pulse, z_i is the number of electrons that would be freed by complete oxidation of the substrate to CO_2, and F is the Faraday constant. Thus, the first term in the denominator is the maximum charge available from the concentration pulse. $\dot{V}$ is the outlet flow rate, c_j^{out} the outlet concentration of substance j, $c_{0,j}^{out}$ is the outlet concentration of substance j in steady state. Thus the second term in the denominator accounts for the additional charge that is flushed out of the reactor above the steady state level and the current efficiency captures only the effects from the pulse, not the steady state. For comparison, the steady state current efficiency is reported in the last line of table 6.2. It is calculated by equation 6.2, considering an average 1,3-propanediol concentration of $0.1\,\mathrm{mmol\,l^{-1}}$, which is found in the effluent stream under closed-circuit conditions:

$$\mathrm{CE_{ss}} = \frac{I_0}{c_{\mathrm{Gly}}^{\mathrm{in}} \dot{V} z_{\mathrm{Gly}} \mathrm{F} - \sum_j \dot{V} c_j^{\mathrm{out}} z_j \mathrm{F}} \tag{6.2}$$

The remaining carbon is either converted to biomass or methane which were not quantified. The good agreement between the CE of 26.7 % from the glycerol concentration pulse and the CE of 28.4 % in steady state indicates that the pathways and mechanisms active during the pulses were similar to those in steady state.

In literature, a CE of 32-35 % has been reported for a glycerol-fed MFC [110]. The conversion efficiency of 66 % for acetate is comparable to values reported for batch MFC experiments using acetate as the sole carbon source [167], which confirms that the biofilm is highly adapted to acetate despite glycerol being supplied at the inflow.

6.3.6 Rate constants in the planktonic phase

In the following section, it will be demonstrated how quantitative information about the reaction rates of glycerol electrooxidation can be extracted from the pulse experiments. Rate constants for the substrate consumption in the liquid phase can be calculated when using the Monod equation

$$r_i = X_j \cdot q_{max,i} \frac{c_i}{c_i + K_{S,i}} \tag{6.3}$$

to describe the substrate consumption rate r_i of substance i. X_j is the biomass in the biofilm (for acetate consumption) or in the planktonic phase, c_i the substrate concentration, $q_{max,i}$ the saturation rate constant and $K_{S,i}$ the half-saturation rate constant. For the time directly after a concentration pulse, it is reasonable to assume $c_i \gg K_{S,i}$ for the pulsed intermediate. Using this assumption, the concentration transients after the pulse are described by a simplified model according to equation 6.4, with the reactor volume V, the flow rate $\dot{V}$, and the concentration c_i of the substance that was injected for the respective pulse experiment.

$$V \cdot \frac{d c_i}{d t} = -\dot{V} \cdot c - X_j \cdot q_{max,i} \tag{6.4}$$

The differential equation was implemented and solved in the software Matlab. The maximal substrate utilization rate $X_j \cdot q_{max,i}$ was adjusted to fit the simulated concentration transients to the first three experimental concentration values after the pulse. The experimental and the simulated values, which are depicted in figure 6.8, agree well for the time shortly after the concentration pulse. This agreement shows that the assumption $c_i \gg K_{S,i}$ is sound. The resulting values for $X_j \cdot q_{max,i}$ can be found in table 6.3. The concentration transients for glycerol and 1,3-propanediol pulses, which were recorded under closed-circuit and open-circuit conditions, can be described by the same maximal substrate utilization rates. This again confirms that the reactions in the planktonic phase were largely independent of the electrode reactions within the limited time of operation at open-circuit conditions. The estimation of net rate constants for the total consumption in the planktonic phase cannot replace carefully designed kinetic experiments for the

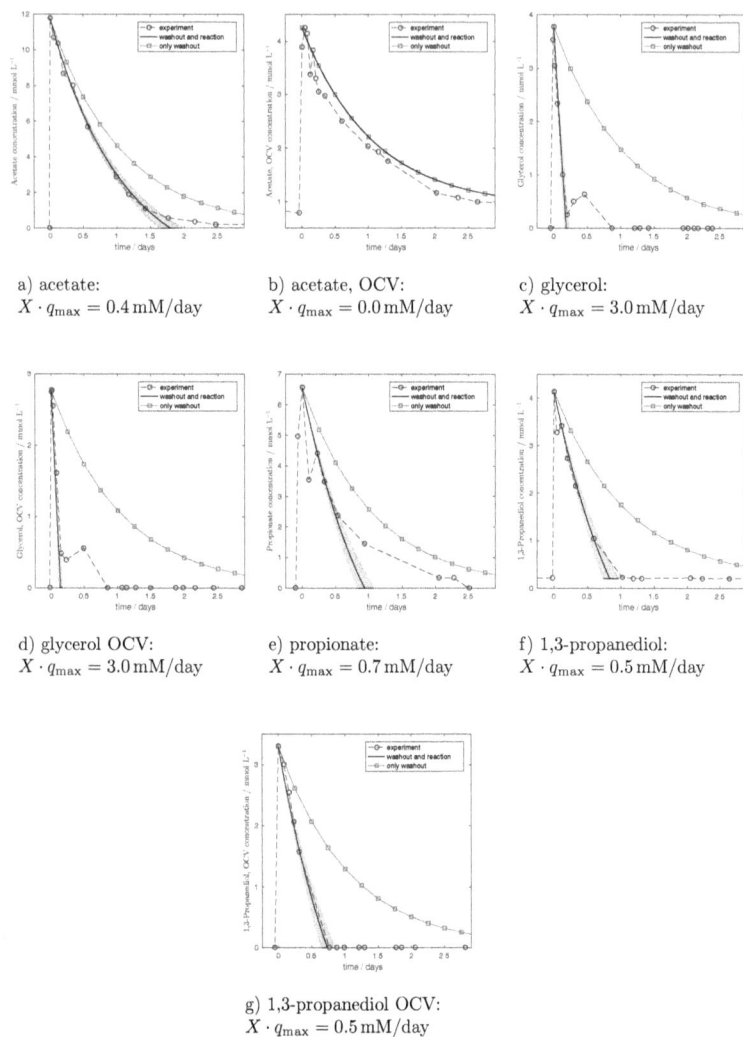

a) acetate:
$X \cdot q_{max} = 0.4\,\mathrm{mM/day}$

b) acetate, OCV:
$X \cdot q_{max} = 0.0\,\mathrm{mM/day}$

c) glycerol:
$X \cdot q_{max} = 3.0\,\mathrm{mM/day}$

d) glycerol OCV:
$X \cdot q_{max} = 3.0\,\mathrm{mM/day}$

e) propionate:
$X \cdot q_{max} = 0.7\,\mathrm{mM/day}$

f) 1,3-propanediol:
$X \cdot q_{max} = 0.5\,\mathrm{mM/day}$

g) 1,3-propanediol OCV:
$X \cdot q_{max} = 0.5\,\mathrm{mM/day}$

Figure 6.8: Experimental and simulated concentration transients for the pulsed component during the pulse experiments. Values of $X \cdot q_{max}$ were adjusted to reproduce dynamic experiments. The gray corridor indicates the concentration transients for the respective value of $X \cdot q_{max} \pm 20\,\%$, the propionate pulse response was not determined under open-circuit conditions. Reprinted from [152], SI (CC BY 4.0).

Table 6.3: Estimated maximal rates of substrate consumption directly after the concentration pulse, assuming $c_i \gg K_{S,i}$ using equation 6.4. For acetate consumption, the biomass X_j refers to the biofilm mass, for all other substrates it refers to the biomass in the planktonic phase. Reprinted from [152] (CC BY 4.0).

injected substrate i	$X_j \cdot q_{max,i}$
acetate	0.4 mmol day^{-1}
1,3-propanediol	0.5 mmol day^{-1}
propionate	0.7 mmol day^{-1}
glycerol	3 mmol day^{-1}

single reactants and yields quite large error ranges. As it can be seen in figure 6.8, changing the rate constants by 20 % still yields a reasonable fit. However, the possibility of obtaining approximate rate constants for various substances in a single continuously operated reactor is a clear advantage from a practical point of view.

6.3.7 Rate constants in the biofilm

Parameters for the electroactive biofilm can also be obtained from the pulse experiments, as will be shown in this section. While pulse experiments have already been applied in anaerobic digestion [168], to the best of the author's knowledge, no rate constants for MFCs have been derived from pulse experiments yet.

The current produced by the biofilm is described by the the Nernst-Monod equation that was introduced in chapter 2:

$$I = zFq_{max,ac}X_{bf}\frac{c}{c + K_S}\frac{1}{1 + \exp\left[-\frac{F}{RT}(E - E_{K_A})\right]} \tag{6.5}$$

In turnover CVs (see figure 6.9), the current reaches a plateau far below the potential of 0.2 V applied in CA, showing that the potential dependent term is very close to one throughout all CA experiments. Directly after the concentration pulse, see figure 6.4 (A), the concentration dependent term is also close to one because of the high acetate concentration. The fact that the current reaches a plateau 22 hours after the pulse and remains nearly constant for seven hours in

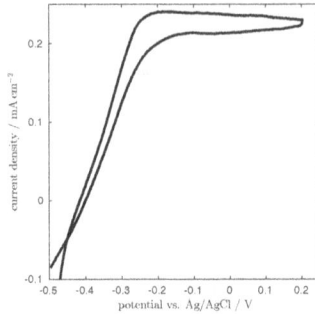

Figure 6.9: Turnover CV recorded before the concentration pulse experiments showing that the current is not limited by the anode potential above -0.1 V. Residence time is 28.8 h, glycerol inlet concentration 1.8 mM, scan rate 1 mV/s. Reprinted from [152], SI (CC BY 4.0).

spite of a continuously falling acetate concentration, is a further indicator that the biofilm current production is not limited significantly by acetate concentration during the whole growth phase that follows the concentration pulse. Therefore the increase in current density is attributed to an increase in $q_{max,ac}X_{bf}$. From the current density directly after the concentration pulse, $q_{max,ac}X_{bf} = 0.164 \,\mathrm{mmol\,d^{-1}}$ can be calculated by Faraday's law. From this value and $q_{max,ac}X_{bf}$ at the end of the growth phase, a rate for the increase of $q_{max,ac}X_{bf}$ over time can be calculated: $\frac{\mathrm{d}\,(q_{max,ac}X_{bf})}{\mathrm{d}t} = 0.560 \,\mathrm{mmol\,d^{-2}}$ which corresponds to $0.560 \,\mathrm{mmol\,m^{-2}\,d^{-2}}$. Even though the growth seems to proceed linearly, it is likely that the observable behaviour results from a number of non-linear processes, including not only an increase in biomass, but also changes in gene expression and increasing transport resistances for substrate and protons in the thicker biofilm.

Approximately 1.5 days after the pulse, the current density begins to fall. In figure 6.10, it is shown that this decline in current density can be described by a Monod-type relationship, $I \propto c/(c + K_s)$, with a half-saturation rate constant $K_S \approx 1.4 \,\mathrm{mmol\,L^{-1}}$ for the biofilm after the growth phase. In literature, K_S values of $0.674 \,\mathrm{mmol\,L^{-1}}$ for a mixed culture biofilm and $0.126 \,\mathrm{mmol\,L^{-1}}$ for a *Geobacter sulfurreducens* biofilm have been reported [163]. Depending on the electron acceptor, K_S values of 0.01 and $0.03 \,\mathrm{mmol\,L^{-1}}$ have been reported for *G. sulfurreducens* in suspended cell culture [169]. Such high deviations between reported values

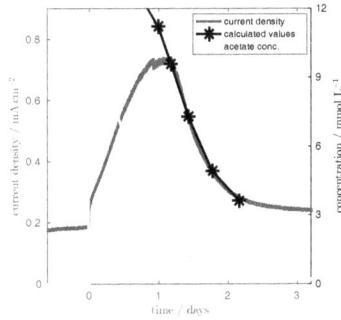

Figure 6.10: Determination of K_S for the biofilm from the acetate concentration pulse. The line with the asterisk markers is calculated by $i = i_{max} \cdot c_{ac}/(c_{ac} + K_S)$ with $K_S = 1.4\,\mathrm{mmol\,L^{-1}}$, and $i_{max} = 1.25\,\mathrm{mA\,cm^{-2}}$. K_S and A were calculated from measured current and concentration values according to: $\frac{i(t_1)}{i(t_2)} = \frac{i_{max} \cdot c_{ac}(t_1)}{(c_{ac}(t_1) + K_S)} \cdot \frac{(c_{ac}(t_2) + K_S)}{i_{max} \cdot c_{ac}(t_2)}$. Reprinted from [152], SI (CC BY 4.0).

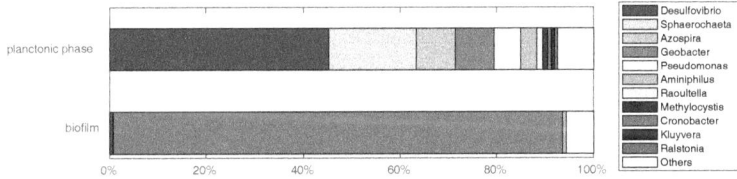

Figure 6.11: Relative abundance of bacterial genera found in biofilm and planktonic phase of the reactor after 2 months of continuous operation with 1.8 mM glycerol inlet concentration, an anode potential of 0.2 V, and a residence time of 28.8 h. Reprinted from [152] (CC BY 4.0).

occur because the apparent half-saturation rate constant in a biofilm strongly depends on transport. The calculations above show that concentration pulse experiments yield information on concentration dependency and growth behaviour of the biofilm. Of course, the parameters of interest will vary according to the biofilm model in use and the specific research question. In principle, concentration and time dependent models for biofilm performance using other relationships than the Nernst-Monod equation can also be parametrised by pulse experiments.

6.3.8 Microbial community analysis

In figure 6.11, the results from a sequencing analysis are presented, which was conducted two months after the start of the experiment. The sequencing analysis provides only a snapshot of the community composition which might be changing over time. However, the fact that the repetition of the glycerol concentration pulse experiment shows the same distinct features as the first glycerol concentration pulse experiment indicates that at least the part of the community that dominates the observable fermentation in the planktonic phase and the current production in the biofilm was stable over time in the system. It can be seen in figure 6.11 that there is a clear division between planktonic phase and biofilm: while the biofilm community is relatively homogeneous and dominated by species from the *Geobacter* genus, the planktonic phase is comparatively heterogeneous. In the biofilm, *Geobacter* spp. accounted for a share of 92.8 %. No other single bacterial genus accounted for more than one percent of the reads. The high share of *Geobacter* spp. most likely results from the inoculation of the reactor with an acetate-grown secondary biofilm [170]. It is remarkable though, that the population is stable after transfer into a glycerol-fed reactor and subsequent operation for two months. After taking the samples for sequencing, the reactor was sealed again, as explained in the experimental section. Following a brief recovery phase, the current density returned to the previous steady state value of approximately $0.2\,\mathrm{mA/cm^2}$.

In the planktonic phase, members of the *Desulfovibrio* genus comprise the largest share (45.2 %), followed by the genera *Sphaerochaeta* (18.1 %), *Azospira* (8.1 %), *Geobacter* (8 %), and *Pseudomonas* (5.6 %).

The presence of *Geobacter* spp. might result from biofilm abrasion. *Desulfovibrio alcoholovorans* and *D. fructovorans* can ferment glycerol to 3-hydroxypropionate and 1,3-propanediol in the absence of electron acceptors [171] and to acetate in the presence of electron acceptors such as sulfate or methanogens [172]. *D. alcoholovorans* is also able to oxidise 1,3-propanediol to acetate in the presence of electron acceptors [172, 171]. Also other species from the *Desulfovibrio* genus can produce acetate from complex substrates. For example, *D. desulfuricans* produces acetate from lactate [173] or polyethylene glycol [174]. The pathways identified from the concentration pulse experiments are thus consistent with literature results. Further investigations of the interactions between the species are necessary to explain the precise reason for the large share of *Desulfovibrio* spp. in the system.

Desulfovibrio was found by [175] in the planktonic phase of an MEC where it was assigned the role of homo-acetogenesis from hydrogen and CO_2. In this work, the increase in acetate concentration under open circuit conditions, when compared to closed-circuit conditions, is consistent with the decrease in current (see section 6.3.3). In conclusion, homo-acetogenesis from hydrogen produced at the cathode did not contribute significantly to acetate production.

The other genera found in the reactor have also been observed previously in BES: *Sphaerochaeta* spp. were observed in a BES for the anaerobic digestion of molasses [176] and closely related *Spirochaeta* spp. were found in MFCs running on wastewater [177]. *Pseudomonas* spp. were found in BES running on acetate [165], and *Azospira* spp. have been found in wastewater-fed BES [178]. But the role of these bacteria has not been clearly identified. [154] reviewed glycerol fermentation to 1,3-propanediol. They concluded that the following bacteria are able to ferment glycerol to 1,3-propanediol: *Klebsiella, Enterobacter, Citrobacter, Lactobacilli,* and *Clostridia.* However, no significant numbers ($< 0.04\%$) of bacteria from these genera were observed in the system.

The low glycerol inlet concentration in combination with the continuous flow operation might have exercised a selective pressure that was not present in other studies, leading to a different microbial community. Additionally, the inoculum from an acetate acclimatised MFC might have lacked certain bacteria in the first place. However, these conditions did not negatively affect the coulombic efficiency as discussed above.

The community analysis confirms the division of labour principle that was already visible in the concentration pulse experiments and suggests that *Desulfovibrio* spp. species might be interesting candidates for designing syntropic communities for BES utilizing glycerol.

6.4 Conclusions for the experimental part

A fast and analytically simple method for identifying reaction pathways in continuously operated glycerol-fed BES under chemostatic conditions by applying concentration pulses was presented. It was found that glycerol is first fermented by planktonic bacteria to acetate which is then oxidised electrochemically by the anode respiring bacteria in the biofilm. Additionally, rate constants for the metabolism of various intermediates, as well as kinetic parameters for the biofilm, were obtained.

Community analysis results agreed with the conclusions from the pulse experiments and confirm the division of labour principle between biofilm and planktonic phase. The presented concentration pulse method can be used to identify reactor behaviour and even evaluate kinetics without community analysis and without a need for starting numerous batch experiments. Unlike the approaches presented in part one of this thesis, the concentration pulse data can be interpreted without a dynamic physical model that requires detailed knowledge of the reaction mechanisms.

In the following sections it will be shown that further insights into the biofilm can be obtained from the set of experimental data with the help of a mathematical model.

6.5 Modelling and parameter identification for the biofilm

The concentration pulse method allowed to identify pathways in the BES and separate processes in the liquid phase and the biofilm. However, the processes that determine the current production and growth rates of the electroactive biofilm cannot be correlated directly to the experimentally observed variables. In general, the *in-operando* characterization of a biofilm anode is a difficult task because, unlike planktonic cells, the microorganisms are firmly attached to the anode and do not leave the reactor with the effluent.

To obtain an better understanding of the biofilm electrode, which is a key component in any BES, a model for the biofilm electrode will be developed and analysed in this section. The model will be used to gain further insights into processes that were not observed directly in the experiments.

6.5.1 Background and approach

Existing MFC models have been thoroughly reviewed in literature, for instance in [35] and [179]. Depending on the specific application, models cover half cells or full cells, consider different processes and possess variable levels of detail.

In an early study, Marcus et al. [30] investigated the influence of various parameters on anodic biofilms based on a sophisticated spatially resolved biofilm model. Such pure modelling studies are important to elucidate fundamental factors influencing the biofilm performance. Another important goal of modelling is the determination of actual parameter values. In several studies, physically meaningful parameters were identified based on experimental data. Sedaqatvand et al. [180] analysed a single-chamber MFC for dairy-wastewater treatment. A sophisticated model for the full cell was established, and six parameters were identified by fitting to experimental data and eight others were taken from literature. Parameter sensitivity and confidence intervals were not considered. Zeng et al. [181] identified six parameters of a model for a two-chamber acetate-fed MFC and fixed four additional parameters to literature values. A one-by-one sensitivity analysis revealed which parameters had the biggest influence on cell performance, but the confidence intervals for these parameters were not analysed.

In general, rigorous parametrization including identifiability analysis and determination of confidence intervals is not common in MFC modelling, even though these techniques are routinely applied to modelling other biological systems with mixed cultures [182, 183], due to the complexity and large number of unknown parameters.

To fill this gap, it will be demonstrated how a rigorous parametrization of a model for a biofilm anode can be achieved. To this end, a dynamic model for a biofilm anode is developed, parameters that reproduce the experimental data are identified, a theoretical identifiability analysis is performed and confidence intervals for the model parameters are determined.

For the anodic biofilm inside the glycerol-fed BES, time dependent data of current and concentration is available. The model, which is set up and parameterised rigorously, describes the electrooxidation of acetate, the growth of the anodic biofilm and the development of the substrate concentration inside the cell. The model is parameterised using the acetate concentration pulse response. During concentration pulse experiments the concentration is expected to cross the value of the half-saturation rate constant. This is an advantage for parameter identification because, as pointed out by Batstone et al., the maximum substrate utilization rates and half-saturation rate constants from the well-known Monod-Kinetic can be differentiated better under these conditions [183]. However, the methodology which is presented can also be used with other sets of experimental data. It will be shown that growth rates and parameters describing the current production and concentration dependency of the MFC anode can be identified with good confidence intervals from concentration pulse response experiments.

In the following section, first the model assumptions and the model equations are discussed. Then the parameter identification procedure is explained, and a theoretical identifiability analysis is performed. The simulation results are compared to the experiments and parameter values and confidence intervals are discussed within the context of the literature.

6.5.2 Modelling

The model developed here does not aim at containing a maximum level of detail but at establishing a simple model with a low number of meaningful parameters that can be reliably identified from the data set. The model is developed in order to reproduce the experimentally observed development of current density over time following a pulse in acetate concentration and in order to describe the underlying processes of biofilm growth and substrate utilization. Only the bioelectrochemical acetate oxidation reaction that takes place inside the biofilm is considered; dynamics are described on the time scale of biofilm growth processes, i.e. hours; and only balance equations for substrate concentration and biofilm biomass are considered. The reactor with volume V is modelled as a continuously stirred tank reactor, i.e. without spatial concentration distribution, which is a reasonable assumption because the biological processes are slow compared to the mixing processes in the stirred reactor. Furthermore, it is assumed that acetate is only consumed through electrooxidation by the biofilm under anaerobic conditions. The acetate concentration balance is given in equation 6.6.

$$\frac{\mathrm{d}\, c_{\mathrm{ac}}}{\mathrm{d}\, t} = \frac{1}{V}\left(r_{\mathrm{p}} - r_{\mathrm{c}} - \dot{V} c_{\mathrm{ac}}\right) \tag{6.6}$$

It is affected by the rate of acetate production from glycerol r_p, the rate of acetate consumption in the biofilm r_c, and the continuous outflow $\dot{V}$ from the reactor. It has been shown above that acetate production from glycerol occurs in the planktonic phase of the reactor whereas acetate electrooxidation takes place in the anode-attached biofilm. Thus, the rate of acetate production from glycerol is treated as a constant parameter. This is reasonable because the glycerol inflow remains constant at all times. The fact that no changes in the concentrations of other intermediates of the glycerol metabolism were detected by HPLC after the acetate concentration pulse indicates that acetate concentrations in the experiment's concentration range did not inhibit glycerol metabolism. The electrochemical oxidation reaction of acetate is:

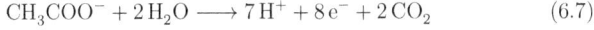

$$CH_3COO^- + 2\,H_2O \longrightarrow 7\,H^+ + 8\,e^- + 2\,CO_2 \tag{6.7}$$

As in section 6.3.7, the rate of acetate consumption by the biofilm is modelled via the Nernst-Monod equation:

$$r_c = q_{max}X_{bf}\frac{c_{ac}}{c_{ac} + K_S}\frac{1}{1 + \exp(-FR^{-1}T^{-1}(E - E_A))} \tag{6.8}$$

It describes the biofilm activity as a function of the maximum substrate utilization rate q_{max}, the half-saturation rate constant K_S, the acetate concentration c_{ac} and the difference between the electrode potential E and the half-saturation potential E_{KA}. Only one type of biomass X_{bf} with averaged properties is considered. Since the community analysis showed more than 90 % *Geobacter* species in the biofilm [152], the biofilm properties are most likely dominated by *Geobacter* species. Thus, identified parameter values for the biofilm will be compared to literature values of *Geobacter*. Additionally, only biomass that is active for bioelectrochemical acetate oxidation is considered, which is a simplification since biofilms also contain inactive cells and extracellular polymeric substances (EPS). Sun et al. [184] showed that not the complete anodic biofilm is viable in a microbial fuel cell and demonstrated that current generation was related to protein content of the biofilm rather than to its total thickness. The equation does not consider transport effects that can play an important role in biofilms. Especially the concentration dependent term $\frac{c_{ac}}{c_{ac}+K_S}$ that works well for bacteria in planktonic phase does not necessarily describe a thick biofilm where cells from different layers are facing different substrate concentrations. Additionally, potential and pH gradients within electroactive biofilms can limit performance [20]. To keep the model simple and to not introduce numerous parameters, the effect of all kinds of transport limitations is described by replacing K_S with an effective half-saturation rate constant $K_{S,eff}$ that changes with the active biomass of the biofilm:

$$K_{S,eff} = K_{S,0} + K_{S,0}X_{bf}^2 \tag{6.9}$$

$K_{S,0}$ is a constant that corresponds to the value of $K_{S,eff}$ at a hypothetical biofilm thickness of zero. Further details about the derivation of this relation can be found in section A.4 in the appendix. The results below will show that it captures the biofilm behaviour with reasonable accuracy. Here, it is assumed that the biofilm grows only in direction perpendicular to the electrode with a constant biofilm density. The film thickness is thus proportional to the biofilm mass. It is assumed that a constant fraction f_0 of the consumed acetate is converted into active biomass with a constant average molar mass M_{bf} that is normalised to the carbon content of the active biomass. The biofilm growth is limited by a decay rate that is proportional to the decay factor k_d and the active biomass X_{bf}. Equation 6.10 describes the mass balance for the biofilm mass:

$$\frac{d\,X_{bf}}{d\,t} = 2r_c f_0 M_{bf} - k_d X_{bf} \tag{6.10}$$

Only metabolically active cells are considered, and all types of bacteria are lumped into a single biofilm mass state variable. In literature, more detailed descriptions have been presented where active and inactive biomass was differentiated [30]. For parameter identification, a simple model with fewer state variables is preferred, as long as the experimental data can be explained. Since the biofilm is firmly attached to the electrode, there is no loss of biofilm due to the continuous flow in the reactor. The decay factor k_d incorporates all effects that reduce the activity of the biomass such as cell death, biofilm abrasion or a reduction in the amount of proteins involved in EET in living cells.

The share of consumed substrate that is not used for cell growth is oxidised electrochemically according to reaction 6.7 with $z = 8$ mol electrons per mol acetate. The resulting external current density i, i.e. current per electrode area A_{el}, is calculated as:

$$i = (1 - f_0) r_c z F A_{el}^{-1} \tag{6.11}$$

The presented model equations describe the dynamics of biofilm growth, acetate consumption and current production that take place over a time scale of days. Therefore, rapid dynamic processes such as double layer charging [28], diffusion and internal substrate storage as described in [164] and [131] are assumed to

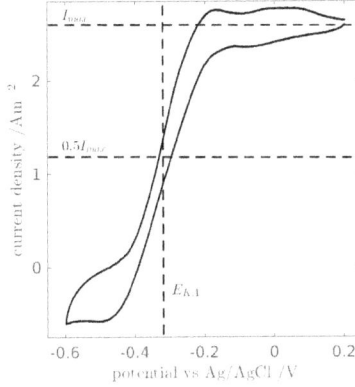

Figure 6.12: Turnover CV at a scan rate of $1 \, \mathrm{mV \, s^{-1}}$. At a potential of $0.2 \, \mathrm{V}$, the current production is not limited by potential. The half-saturation potential E_{KA} is $-0.35 \, \mathrm{V}$.

be in quasi steady state. In total, the model contains two state variables, X_{Bf} and c_{ac}, and seven unknown parameters: f_0, k_d, $K_{S,0}$, q_{max}, r_p, E_{KA}, and M_{bf}. Values of A_{el}, $\dot{V}$, V and T are set to their respective experimental values. In the following sections, the identification of the unknown parameter values and the determination of their confidence intervals will be discussed.

6.5.3 Parameter identification

The half-saturation potential E_{KA} that describes the dependency of the reaction rate on the electrode potential can be directly determined from a CV. The turnover-CV, from which $E_{KA} = -0.35 \, \mathrm{V}$ is obtained, can be seen in figure 6.12. The molar mass of the biofilm per carbon atom is fixed to $M_{bf} = 24 \, \mathrm{g \, mol^{-1}}$ based on literature [185]. Table 6.4 lists the values of all fixed model parameters, including experimental conditions and constants.

The remaining five model parameters f_0, k_d, $K_{S,0}$, q_{max} and r_p are identified from the experimental concentration and current transients from the acetate concentration pulse described above. A log-likelihood function that describes the logarithm of the probability of observing the experimental data for a given set of model parameters Θ based on the deviation between experiment and simulation

Table 6.4: Fixed model parameters

Parameter	Value	Notes
A /m^2	$6 \cdot 10^{-4}$	From experiment
F /C mol^{-1}	96485	
R /J mol^{-1} K^{-1}	8.3144	
T /K	308.15	From experiment
$\dot{V}$ /L s^{-1}	$1.73 \cdot 10^{-6}$	From experiment
V /L	0.180	From experiment
M_{bf} /kg mol^{-1}	0.0246	Assumed
E_{KA} /V	-0.35	From turnover CV

and the standard deviation of the measurements is used. It is maximised to find the best parameter set:

$$
\text{LogL}(\Theta) = -\frac{n_i}{2}\ln(2\pi\sigma_i^2) - \frac{1}{\sigma_i^2}\sum_{j=1}^{n_i}\left(i_j^{\text{sim}} - i_j^{\text{exp}}\right)^2
$$
$$
- \frac{n_{\text{conc}}}{2}\ln(2\pi\sigma_{\text{conc}}^2) - \frac{1}{\sigma_{\text{conc}}^2}\sum_{k=1}^{n_{\text{conc}}}\left(c_k^{\text{sim}} - c_k^{\text{exp}}\right)^2 \tag{6.12}
$$

For the n_{conc} concentration data points, the standard deviation σ_{conc} is $0.1\,\text{mol}\,\text{m}^{-3}$ and for the n_i current density data points the standard deviation σ_i is $0.06\,\text{A}\,\text{m}^{-2}$ here. This formulation of the objective function can be easily adjusted to incorporate different measured variables with varying sampling rate without a need for further scaling factors.

The model equations were implemented in the software MATLAB and the objective function was maximised using a Lewis and Torczon generalised pattern search algorithm [186] as implemented in MATLAB's global optimization toolbox.

The upper and lower boundaries for the parameter values given in Table 6.5 were based on physical considerations, where possible, or otherwise assumed. In particular, f_0 must be larger than 0 and smaller than 1 and the acetate production rate r_p is limited by the total carbon supply from glycerol inflow.

Table 6.5: Optimal values and confidence intervals of the identified parameters based on profile likelihood calculations

Parameter	99% confidence interval	95% confidence interval	Lower and upper boundary	[30]
q_{max} /mol s^{-1} kg^{-1}	$[2.31 \cdot 10^{-3},$ $2.51 \cdot 10^{-3}]$	$[2.33 \cdot 10^{-3},$ $2.48 \cdot 10^{-3}]$	$[0, 0.01]$	$1.53 \cdot 10^{-3}$
$K_{S,0}$ /mol m^{-3}	$[0.03, 0.037]$	$[0.03, 0.036]$	$[0, 0.2]$	0.03
f_0 /-	$[0.431, 0.450]$	$[0.443, 0.448]$	$[0, 1]$	$0.05 - 0.1$
k_d / s^{-1}	$[3.33 \cdot 10^{-5},$ $3.73 \cdot 10^{-5}]$	$[3.37 \cdot 10^{-5},$ $3.68 \cdot 10^{-5}]$	$[0, 1 \cdot 10^{-4}]$	$[5.8 \cdot 10^{-7},$ $2.31 \cdot 10^{-6}]$
r_p / mol s^{-1}	$[2.57 \cdot 10^{-9},$ $2.70 \cdot 10^{-9}]$	$[2.59 \cdot 10^{-9},$ $2.69 \cdot 10^{-9}]$	$[0, 3.2 \cdot 10^{-9}]$	-

6.5.4 Structural identifiability analysis

The fact that a parameter set exists that describes the experimental data reasonably well does not necessarily mean that the parameter values are accurate and reliable. This is because any set of experimental data can be described with an arbitrarily complex model and a sufficiently large number of parameters.

Thus, first a structural identifiability analysis is conducted to explore if the experimental data contains sufficient information to uniquely identify all model parameters. While there are numerous methods published in literature to check for identifiability, the recently published approach by Stigter el al. [187] is chosen, which is based on a singular value decomposition of the sensitivity matrix S [188, 189] of the model. This matrix describes the sensitivity of the measured quantities, here current density and acetate concentration, towards the model parameters Θ for every point in time:

$$S(t, \Theta) = \mu_1 \zeta_1 v_1 + ... + \mu_N \zeta_N v_N \tag{6.13}$$

If one of the N singular values ζ_i determined by the singular value decomposition in equation 6.13 is zero, within numerical accuracy, the model is structurally non-identifiable. The numerical calculation of the singular value decomposition and a geometrical interpretation of the singular values can be found in [190].

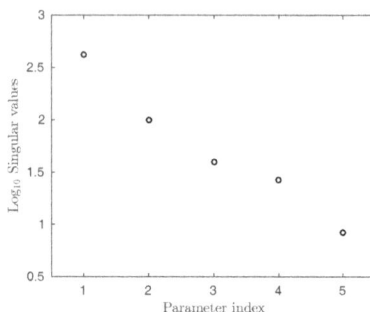

Figure 6.13: Five singular values obtained from the singular value decomposition of the model's sensitivity matrix. All values are larger than zero, showing that the parameters are structurally identifiable.

6.5.5 Results and discussion

6.5.5.1 Structural identifiability

A parameter set was identified that allows reproduction of the experimental data. The resulting parameter values and the simulation results are discussed in the following section 6.5.5.3. The sensitivity matrix was obtained during solving the model's differential equations and a singular value decomposition was performed as described above. In figure 6.13, the singular values for the investigated system are plotted. All singular values are clearly greater than zero. This means that the system is locally identifiable, and all parameter values can be identified unambiguously if the accuracy of the measurement is sufficiently high.

6.5.5.2 Model-supported analysis of biofilm dynamics

In the upper panel of figure 6.14, the simulation result for the identified parameter set is compared to the experimental current and concentration data.

In the experiment, the current jumps almost instantly from 1.8 to 2.4 A m^{-2} upon the concentration pulse. Following the initial jump, the current density increases linearly over time, then slows down, reaching a plateau of 7.3 A m^{-2} after one day. From 1.25 days on, the current density falls quickly. Acetate concentration decreases monotonously after the acetate pulse.

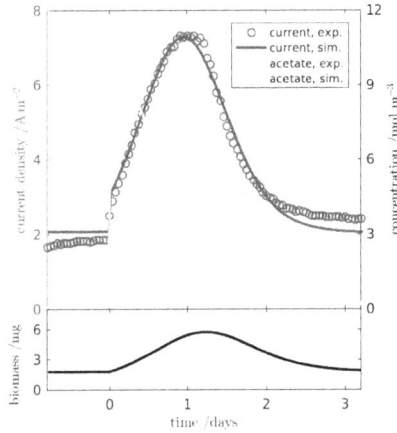

Figure 6.14: Comparison of experimental and simulated concentration and current response of the biofilm to an acetate concentration pulse at t=0. Anode potential is 0.2 V, residence time is 28.8 hours.

The simulation quantitatively reproduces the jump in current density directly after the pulse, as well as the subsequent linear increase of current density. The plateau of the current density and the subsequent decline can also be described by the model. The simulated concentrations follow the experimental curve with minor deviations.

The jump in current density upon the pulse is attributed to the concentration-dependent Monod-term $c_{ac}/(c_{ac} + K_S)$ of the Nernst-Monod kinetic (equation 6.8) which reacts without a time delay whereas the biomass growth takes rather hours and days. The nearly linear increase of current density over time after the pulse can be explained by an increase in active biofilm mass. In the lower panel of figure 6.14, the simulated development of biofilm mass over time is displayed. The active biofilm mass increases from 1.6 mg in steady state to a maximum of 6.2 mg before starting to fall again. X_{bf} is only biomass that is active for current generation. The decline in X_{bf} after the pulse does not necessarily go hand in hand with a proportional decline in geometrical biofilm thickness because inactive components and EPS matrix might remain intact. Compared to steady state conditions, the simulated biofilm mass is about four times larger and acetate concentration is approximately ten times larger when the peak current density is reached. Nevertheless, the current density increases only four-fold. The higher acetate concentration is partly compensated by the increasing transport resistance of the thicker biofilm. Current densities reported in literature usually do not exceed $10\,A\,m^{-2}$ of geometric surface area [191] regardless of substrate concentration, which is consistent with additional limitation occurring in thicker biofilms.

Assuming a biofilm density ρ_{Bf} of $40\text{-}200\,mg\,cm^{-3}$ [30] with an electrode area A_{el} of $6\,cm^2$, the steady state active biofilm mass of $1.6\,mg$ corresponds to a biofilm thickness of $\delta_{bf} = X_{bf}/(A_{el}\rho_{bf})$ in the range of $13.3\text{-}66.6\,\mu m$. This is a lower boundary estimate because inactive biomass is not considered. Compared to literature where biofilms of $110\,\mu m$ [19] and $20\text{-}40\,\mu m$ [184] thickness have been found on graphite electrodes, it is a realistic value. All in all, the simulation results agree well with the experimental data and with published results on biofilm thickness and transport resistances.

6.5.5.3 Parameters and parameter correlations

To evaluate if the identified parameter values are reliable, the parameter values and confidence intervals will be discussed and compared to literature in this section. In table 6.5, the values of the parameters and their $95\,\%$ confidence intervals are reported.

In comparison to literature values, the maximum substrate utilization rate q_{max} of $0.0024\,mol\,s^{-1}\,kg^{-1}$ is very close to the value of $0.0023\,mol\,s^{-1}\,kg^{-1}$ used by Picioreanu et al. [192], and moderately higher than the $0.0015\,mol\,s^{-1}\,kg^{-1}$ reported by Marcus et al. [30] for *Geobacter sulfurreducens*.

The half-saturation rate constant $K_{S,0} = 0.033\,mol\,m^{-3}$ is very close to the value of $0.03\,mol\,m^{-3}$ used by Marcus et al. [30] for *Geobacter sulfurreducens*. The effective half-saturation rate constant in steady state of $K_{S,eff} = 0.12\,mol\,m^{-3}$, which was calculated according to equation 6.9 with a steady state biofilm mass of $1.6\,mg$, is significantly higher, which is reasonable because the value already incorporates transport limitations.

The share of carbon utilised for cell growth of $f_0 = 0.44$ as well as the biomass decline factor k_d of $3.53 \times 10^{-5}\,s^{-1}$ are both very high compared to literature values. Marcus et al. [30] use f_0 values ranging from 0.05 to 0.1 and k_d ranging from $1.15 \times 10^{-6}\,s^{-1}$, calculated from the rates of endogenous respiration and biomass deactivation, to $4.45 \times 10^{-6}\,s^{-1}$, calculated from the rates of endogenous respiration, biomass deactivation and biofilm detachment. The high value of k_d might result from the fact that in the experiment the reactor was stirred at a comparatively high rate of approximately 100 rpm that causes a high shear force and increases biofilm abrasion [193]. An additional factor could be the distribution of active and inactive biomass in the biofilm: Sun et al. [184] found that only the outer layer of an anodic biofilm was viable whereas the inner layer consisted of inactive cells. Since abrasion takes place only in the outer layer, the active biomass might be subject to faster abrasion than the overall biofilm. Wide ranges of f_0 from below 1 % up to 97 % are reported in literature [194] and the values found here are well within range. Future studies may refine the model further. Values of f_0 and k_d that depend on substrate concentration or current density could be included to extend the model, especially if start up phases should be simulated.

Generally, kinetic parameters for mixed cultures should be compared very carefully across different systems because they are based on different model assumption and different experimental conditions. The phenomena and equations that are included in the model and the parameters taken from literature must be chosen carefully according to the questions that should be addressed by the model. This work focuses on describing the development of current generation and substrate consumption over time and does not aim at investigating structural properties of the biofilm or resolving profiles state variables such as biofilm density, share of active and inactive biomass, pH, or potential inside the biofilm.

However, the agreement of the parameter values with literature is all in all very satisfactory, considering the fact that a single experiment was sufficient to identify the values and considering that the only information from literature necessary for the model is the average molar mass of the biomass.

The MATLAB toolbox PESTO [195] was used to determine parameter confidence intervals based on profile likelihoods calculated by equation 6.12. The profile likelihoods are obtained by slightly changing one parameter from the optimum and re-optimizing the other parameters. This way, parameter correlations are obtained by plotting the values of re-optimised parameter values against each other. In parallel, Markov Chain Monte Carlo simulations were conducted with a Metropolis-Hastings algorithm [196] that is also implemented in the PESTO toolbox. The results did not diverge from the profile likelihood calculations. With Markov Chain Monte Carlo simulations, a large number of discrete samples is drawn from an unknown distribution in a way that the density of the samples resembles that of the original distribution.

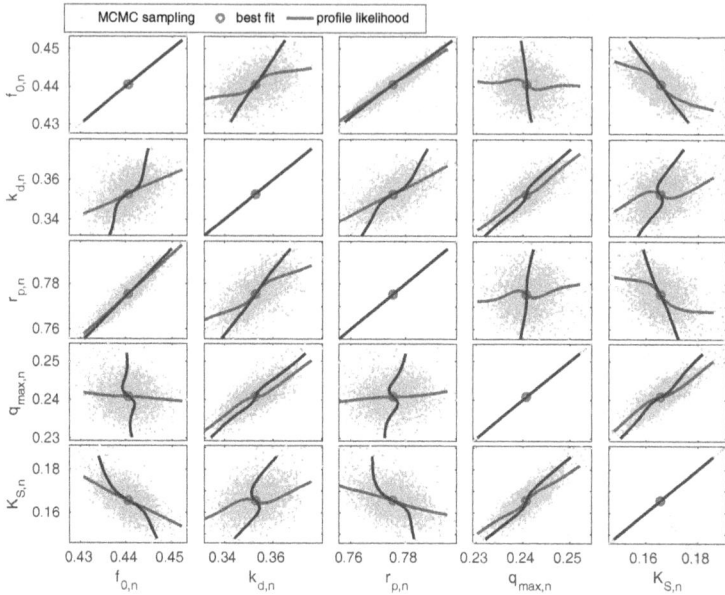

Figure 6.15: Parameter correlations based on profile likelihood calculations (red lines) and Markov Chain Monte Carlo simulation (grey dots). Parameters are normalised between their respective upper and lower boundaries. r_p and q_{max} as well as f_0 and q_{max} are uncorrelated whereas r_p and f_0, q_{max} and $K_{S,0}$ as well as q_{max} and k_d are strongly correlated. The other parameters show an intermediate degree of pairwise correlation.

Figure 6.15 shows the pair-wise parameter correlations based on the profile likelihoods and the Monte Carlo simulations. Note that the parameter values are normalised between their respective upper and lower boundaries in figure 6.15. The correlations calculated from both methods agree with each other. The data shows that r_p and q_{max}, as well as f_0 and q_{max}, are uncorrelated whereas r_p and f_0, q_{max} and $K_{S,0}$, as well as q_{max} and k_d are strongly correlated. The remaining parameters are weakly correlated. Due to the high correlation between certain parameters, linear uncorrelated confidence intervals would not be sufficient to describe the parameter uncertainties.

The correlations can be interpreted physically. For instance, acetate consumption yields either CO_2 and current or biomass. Thus, for given experimental values of acetate concentration and current density, a higher acetate production rate r_p must correspond to a higher value of f_0 to increase biomass production and obey mass conservation. The positive correlation of maximum growth rate q_{max} and the biomass decay coefficient k_d is also reasonable because a steady state biofilm thickness requires growth and decay to be in equilibrium. The positive correlation of q_{max} and $K_{S,0}$ probably results from the fact that acetate concentration is not significantly larger than $K_{S,0}$ most of the time, indicating that design of experiment techniques [197] might help to obtain more informative experiments. As discussed in the introduction, the pulse method still allows one to differentiate q_{max} and $K_{S,0}$ comparatively well. The relative size of the confidence intervals is in a similar range as reported for anaerobic digestion models parametrised with concentration pulse experiments by Kalfas et al. [136] and Batstone et al. [183] indicating that concentration pulses are a suitable tool for identifying parameters for MFCs.

6.6 Conclusions for the modelling part

A methodology to obtain reliable parameters for a physical model of a biofilm anode has been presented. The model in the case study describes substrate consumption rate, biomass growth and current production. The dynamic response of the biofilm to a pulse in substrate concentration could be reproduced well with the model. The simplifying model assumptions – lumping all types of biomass into a single state variable; constant acetate production rate from glycerol; modelling transport resistance in the biofilm by making K_S dependent on biofilm mass – thus seem to be reasonable. Furthermore the application of comparatively simple and transferable methods was demonstrated to show - theoretically and practically - that the pulse experiment contains sufficient information to identify five model parameters with comparatively narrow confidence intervals from a single data set. In conclusion, the concentration pulse experiments are thus a suitable tool to identify model parameters when limited experimental capacities are available. In future, the methodology may be extended towards using input signals of different shapes or a series of concentration pulses. Applying a series of concentration pulses could help to further investigate long term biofilm growth and decay by comparing the successive pulse responses. Applying concentration pulses of different height could help to investigate the concentration dependency of the current production. It is expected that the methodology will be particularly useful for deriving parameters for model-based control or process monitoring but also for evaluating the reliability of information about physical system behaviour obtained from parameter identification.

Chapter 7

Conclusions

In this chapter, first the main findings of the thesis are briefly summarised. Next, the perspectives for future applications of the developed methods are discussed, and some reactions for which these methods are particularly suitable are specified. Finally, promising extensions of the methodological approach are described which include experimental and modelling techniques.

7.1 Summary

The aim of this thesis was to develop and apply new methodologies to determine electrochemical reaction kinetics using dynamic methods, simulations, and online concentration measurements.

To lay the basis for this task, first the most important relations that govern the kinetics in electrochemical cells and in bioelectrochemical systems were discussed.

The first new methodological approach was the combination of DEMS data and electrochemical data to parametrise a physical model of the CO electrooxidation. In chapter 3 a new DEMS cyclone flow cell was presented and characterised that featured well-defined mass transfer to the electrode as well as short response times. The oxidation of CO on a porous Pt/Ru catalyst in the cell was analysed using potential step experiments. The characteristic responses of current density and CO_2 production rate allowed to identify rate constants and transport coefficients using just one set of experimental data. Additionally, it was demonstrated how to model transport of volatile species from the catalyst into the DEMS which is important to estimate time constants.

In chapter 4 the methodology of combining electrochemical data and DEMS data from dynamic experiments with physical models was extended towards a more complex electrochemical reaction, the methanol electrooxidation. With flux-based FRA a new dynamic technique was introduced. It allows to evaluate the production of volatile species as a function of the electrical current in the frequency domain. Kinetic parameters were successfully determined that could describe current and potential as well as the CO_2 production rate for a flux-based FRA spectrum as well as an MSCV with a single set of parameters. Model-based analysis showed that meaningful FRA data can be collected up to a frequency of 0.5 Hz. Furthermore the contributions of transport and reaction to the transfer function were investigated as basis for the future application of the technique.

The second part of the thesis covers bioelectrochemical reactions that feature a higher degree of complexity and a lack of mechanistic data that can be used as a foundation for kinetic analysis.

The first DEMS results on electrochemically active biofilms were presented in chapter 5. The MSCVs revealed that the biofilm was able to oxidise acetate even at a potential a few mV more positive than the standard potential of acetate oxidation. Also storage mechanisms for charge as well as substrate were found and quantified. When no oxidation took place, the biofilm could take up and store substrate at 20 % of the uptake rate under substrate oxidation conditions.

In chapter 6, conversion pathways and rates during bioelectrochemical glycerol oxidation were investigated. In the first part of chapter 6 mainly experimental results were discussed. The system's steady state was distorted for a limited period of time by concentration pulses of individual intermediates. The responses of current density and concentration were monitored using HPLC. This allowed to study the role of the intermediates in the overall reaction and to extract conversion rates of each intermediate. It could be shown that glycerol is first fermented to acetate and then oxidised electrochemically in a biofilm electrode. The electrochemical oxidation was investigated in detail in the second part of the chapter. A model was developed and parameterised that describes dynamics of biofilm growth and acetate oxidation. The advantages of combining electrochemical data and concentration data to determine parameters of physical models were already demonstrated for the two non-biological reactions in the first part of the thesis. In this chapter, a rigorous identifiability analysis and determination of parameter confidence intervals show quantitatively that small confidence intervals can be obtained using just a single dynamic concentration pulse experiment by the inclusion of concentration data.

7.2 Perspectives

The methodologies developed within this thesis provide a toolbox for dynamic analysis of kinetics in electrochemical systems of various degrees of complexity. Some of the approaches and techniques were employed for the first time, such as flux-based FRA using DEMS and the physical modelling of reactions in a DEMS cell, or applied for the first time in a specific field, such as DEMS measurements on electroactive biofilms and the identifiability analysis of a BES model. In general it can be concluded from the results of this thesis that the following factors are beneficial for the identification of electrochemical kinetics:

- application of dynamic techniques such as potential steps, CVs, and EIS because they allow to separate processes by their time constants.

- additional measurements apart from current and potential because transport processes and chemical reaction steps might only have an indirect influence on current and potential.

- physical simulation models that include mass transfer and – if applicable – concentration and potential gradients over the electrode because they allow to access parameters that cannot be directly deduced from experimental data.

- a detailed analysis of the electrochemical cell, that might include CFD analysis, because this information is often necessary to evaluate mass transfer effects.

The techniques introduced in this thesis may also be transferred to other reaction systems. A wide applicability of the developed dynamic approaches was not an explicit goal of this thesis. But the fact that they could successfully be applied to reactions ranging from the well-researched CO electrooxidation, where time constants are in the order of seconds, to highly complex BES, where time constants are in the order of days, indicates that the coupling of dynamic techniques, concentration measurements and simulations is indeed promising.

Flux-based FRA is particularly suitable for reactions that include multiple volatile reactants such as electrochemical CO_2 reduction where CO_2, H_2, and C_2H_4 may be detected or electrochemical ammonia synthesis where N_2, NH_3, and H_2 may be detected. The application of DEMS for the study of electroactive biofilms also opens up opportunities for future research. Especially an extension of the presented experiments to pure cultures might enable a deeper understanding of EET mechanisms in individual organisms. In this context, DEMS might also be used to analyse oxygen reduction in cathodic biofilms or hydrogen consumption in anodic biofilms.

The experimental methods and the modelling techniques applied in this thesis are by no means comprehensive and other methods may be used in future. Thus the overall approach of coupling dynamic techniques, online concentration measurements, and simulations may be extended to further experimental and simulation techniques.

The concentration measurements by DEMS and HPLC can only detect species from the solution and no adsorbed species. When studying non-biological catalysts, they need to be complemented by surface sensitive techniques such as Raman spectroscopy, XANES, or FTIR spectroscopy to obtain a more complete picture of the elementary reaction steps. The dynamic models used for parameter identification were all developed from a vast set of literature data to describe one specific reaction. In future, a generic framework that enables the identification of kinetic models rather than just parameters for given models would be valuable to close the gap between black box models such as equivalent circuit models and physical models.

Bibliography

[1] U. Krewer, M. Christov, T. Vidaković, K. Sundmacher, Impedance spectroscopic analysis of the electrochemical methanol oxidation kinetics, J. Electroanal. Chem. 589 (1) (2006) 148–159. doi:10.1016/j.jelechem.2006.01.027.

[2] Q. Mao, U. Krewer, Total harmonic distortion analysis of oxygen reduction reaction in proton exchange membrane fuel cells, Electrochim. Acta 103 (2013) 188–198. doi:10.1016/j.electacta.2013.03.194.

[3] U. Krewer, A. Kamat, K. Sundmacher, Understanding the dynamic behaviour of direct methanol fuel cells: Response to step changes in cell current, J. Electroanal. Chem. 609 (2) (2007) 105–119. doi:10.1016/j.jelechem.2007.06.015.

[4] U. Krewer, M. Pfafferodt, a. Kamat, D. F. Menendez, K. Sundmacher, Hydrodynamic characterisation and modelling of anode flow fields of Direct Methanol Fuel Cells, Chem. Eng. J. 126 (2-3) (2007) 87–102. doi:10.1016/j.cej.2006.09.001.

[5] G. Hinds, In situ diagnostics for polymer electrolyte membrane fuel cells, Curr. Opin. Electrochem. 5 (1) (2017) 11–19. doi:10.1016/j.coelec.2017.08.010.

[6] Y.-W. Choi, H. Mistry, B. Roldan Cuenya, New insights into working nanostructured electrocatalysts through operando spectroscopy and microscopy, Curr. Opin. Electrochem. 1 (1) (2017) 95–103. doi:10.1016/j.coelec.2017.01.004.

[7] J. Larminie, A. Dicks, Fuel Cell Systems Explained, John Wiley & Sons, Chicheste, 2003.

[8] R. P. O'Hayre, S.-W. Cha, W. G. Colella, F. B. Prinz, Fuel Cell Fundamentals, 2nd Edition, John Wiley & Sons, New York, 2009.

[9] U. Krewer, T. Vidaković-Koch, L. Rihko-Struckmann, Electrochemical Oxidation of Carbon-Containing Fuels and Their Dynamics in Low-Temperature Fuel Cells, ChemPhysChem 12 (14) (2011) 2518–2544. doi:10.1002/cphc.201100095.

[10] L. C. Pérez, P. Koski, J. Ihonen, J. M. Sousa, A. Mendes, Effect of fuel utilization on the carbon monoxide poisoning dynamics of Polymer Electrolyte Membrane Fuel Cells, J. Power Sources 258 (2014) 122–128. doi:10.1016/j.jpowsour.2014.02.016.

[11] A. J. Bard, L. R. Faulkner, Electrochemical Methods Fundamentals and Applications, 2nd Edition, John Wiley & Sons, Inc., New York, 2001.

[12] V. Bagotsky, Fundamentals of Electrochemistry, 2nd Edition, John Wiley & Sons, Hoboken, NJ, USA, 2006.

[13] J. Chun, J. H. Chun, Review on the Determination of Frumkin, Langmuir, and Temkin Adsorption Isotherms at Electrode/Solution Interfaces Using the Phase-Shift Method and Correlation Constants, Korean Chem. Eng. Res. 54 (6) (2016) 734–745. doi:10.9713/kcer.2016.54.6.734.

[14] Q. Wen, H. Zhang, Z. Chen, Y. Li, J. Nan, Y. Feng, Using bacterial catalyst in the cathode of microbial desalination cell to improve wastewater treatment and desalination, Bioresour. Technol. 125 (2012) 108–113. doi:10.1016/j.biortech.2012.08.140.

[15] A. Ter Heijne, O. Schaetzle, S. Gimenez, F. Fabregat-Santiago, J. Bisquert, D. P. B. T. B. Strik, F. Barrière, C. J. N. Buisman, H. V. M. Hamelers, Identifying charge and mass transfer resistances of an oxygen reducing biocathode, Energy Environ. Sci. 4 (12) (2011) 5035. doi:10.1039/c1ee02131a.

[16] C. Santoro, C. Arbizzani, B. Erable, I. Ieropoulos, Microbial fuel cells: From fundamentals to applications. A review, J. Power Sources 356 (2017) 225–244. doi:10.1016/j.jpowsour.2017.03.109.

[17] A. Kadier, Y. Simayi, P. Abdeshahian, N. F. Azman, K. Chandrasekhar, M. S. Kalil, A comprehensive review of microbial electrolysis cells (MEC) reactor designs and configurations for sustainable hydrogen gas production, Alexandria Eng. J. 55 (1) (2016) 427–443. doi:10.1016/j.aej.2015.10.008.

[18] S. Sevda, H. Yuan, Z. He, I. M. Abu-Reesh, Microbial desalination cells as a versatile technology: Functions, optimization and prospective, Desalination 371 (2015) 9–17. doi:10.1016/j.desal.2015.05.021.

[19] A. Baudler, I. Schmidt, M. Langner, A. Greiner, U. Schröder, Does it have to be carbon? Metal anodes in microbial fuel cells and related bioelectrochemical systems, Energy Environ. Sci. 8 (7) (2015) 2048–2055. doi:10.1039/C5EE00866B.

[20] S. C. Popat, C. I. Torres, Critical transport rates that limit the performance of microbial electrochemistry technologies, Bioresour. Technol. 215 (2016) 265–273. doi:10.1016/j.biortech.2016.04.136.

[21] A. J. Hutchinson, J. C. Tokash, B. E. Logan, Analysis of carbon fiber brush loading in anodes on startup and performance of microbial fuel cells, J. Power Sources 196 (22) (2011) 9213–9219. doi:10.1016/j.jpowsour.2011.07.040.

[22] V. Lanas, Y. Ahn, B. E. Logan, Effects of carbon brush anode size and loading on microbial fuel cell performance in batch and continuous mode, J. Power Sources 247 (2014) 228–234. doi:10.1016/j.jpowsour.2013.08.110.

[23] R. A. Rozendal, H. V. Hamelers, R. J. Molenkamp, C. J. Buisman, Performance of single chamber biocatalyzed electrolysis with different types of ion exchange membranes, Water Res. 41 (9) (2007) 1984–1994. doi:10.1016/j.watres.2007.01.019.

[24] J. T. Babauta, H. Beyenal, D. A. Boyd, O. Bretschger, B. Chadwick, J. S. Erickson, F. Fabregat-Santiago, Biofilms in Bioelectrochemical System, 1st Edition, no. 1, Wiley, Hoboken, NJ, USA, 2015.

[25] D. R. Lovley, Electromicrobiology, Annu. Rev. Microbiol. 66 (1) (2012) 391–409. doi:10.1146/annurev-micro-092611-150104.

[26] K. Fricke, F. Harnisch, U. Schröder, On the use of cyclic voltammetry for the study of anodic electron transfer in microbial fuel cells, Energy Environ. Sci. 1 (1) (2008) 144. doi:10.1039/b802363h.

[27] F. Harnisch, S. Freguia, A Basic Tutorial on Cyclic Voltammetry for the Investigation of Electroactive Microbial Biofilms, Chem. - An Asian J. 7 (3) (2012) 466–475. doi:10.1002/asia.201100740.

[28] J. Madjarov, S. C. Popat, J. Erben, A. Götze, R. Zengerle, S. Kerzenmacher, Revisiting methods to characterize bioelectrochemical systems: The influence of uncompensated resistance (iR-drop), double layer capacitance, and junction potential, J. Power Sources 356 (2017) 408–418. doi:10.1016/j.jpowsour.2017.03.033.

[29] R. A. Yoho, S. C. Popat, C. I. Torres, Dynamic Potential-Dependent Electron Transport Pathway Shifts in Anode Biofilms of Geobacter sulfurreducens, ChemSusChem 7 (12) (2014) 3413–3419. doi:10.1002/cssc.201402589.

[30] A. Kato Marcus, C. I. Torres, B. E. Rittmann, Conduction-based modeling of the biofilm anode of a microbial fuel cell, Biotechnol. Bioeng. 98 (6) (2007) 1171–1182. arXiv:bit.20858, doi:10.1002/bit.21533.

[31] A. K. Marcus, C. I. Torres, B. E. Rittmann, Evaluating the impacts of migration in the biofilm anode using the model PCBIOFILM, Electrochim. Acta 55 (23) (2010) 6964–6972. doi:10.1016/j.electacta.2010.06.061.

[32] A. K. Marcus, C. I. Torres, B. E. Rittmann, Analysis of a microbial electrochemical cell using the proton condition in biofilm (PCBIOFILM) model, Bioresour. Technol. 102 (1) (2011) 253–262. doi:10.1016/j.biortech.2010.03.100.

[33] M. Rimboud, E. Desmond-Le Quemener, B. Erable, T. Bouchez, A. Bergel, Multi-system Nernst-Michaelis-Menten model applied to bioanodes formed from sewage sludge, Bioresour. Technol. 195 (2015) 162–169. doi:10.1016/j.biortech.2015.05.069.

[34] H. V. M. Hamelers, A. ter Heijne, N. Stein, R. A. Rozendal, C. J. N. Buisman, Butler-Volmer-Monod model for describing bioanode polarization curves, Bioresour. Technol. 102 (1) (2011) 381–387. doi:10.1016/j.biortech.2010.06.156.

[35] C. Xia, D. Zhang, W. Pedrycz, Y. Zhu, Y. Guo, Models for Microbial Fuel Cells: A critical review, J. Power Sources 373 (October 2017) (2018) 119–131. doi:10.1016/j.jpowsour.2017.11.001.

[36] D. Polcari, P. Dauphin-Ducharme, J. Mauzeroll, Scanning Electrochemical Microscopy: A Comprehensive Review of Experimental Parameters from 1989 to 2015, Chem. Rev. 116 (22) (2016) 13234–13278. doi:10.1021/acs.chemrev.6b00067.

[37] D. A. Buttry, M. D. Ward, Measurement of interfacial processes at electrode surfaces with the electrochemical quartz crystal microbalance, Chem. Rev. 92 (6) (1992) 1355–1379. doi:10.1021/cr00014a006.

[38] J.-Y. Ye, Y.-X. Jiang, T. Sheng, S.-G. Sun, In-situ FTIR spectroscopic studies of electrocatalytic reactions and processes, Nano Energy 29 (2016) 414–427. doi:10.1016/j.nanoen.2016.06.023.

[39] H. Wang, Y.-W. Zhou, W.-B. Cai, Recent applications of in situ ATR-IR spectroscopy in interfacial electrochemistry, Curr. Opin. Electrochem. 1 (1) (2017) 73–79. doi:10.1016/j.coelec.2017.01.008.

[40] D.-Y. Wu, J.-F. Li, B. Ren, Z.-Q. Tian, Electrochemical surface-enhanced Raman spectroscopy of nanostructures, Chem. Soc. Rev. 37 (5) (2008) 1025. doi:10.1039/b707872m.

[41] S. Bruckenstein, G. A. Feldman, Radial transport times at rotating ring-disk electrodes. Limitations on the detection of electrode intermediates undergoing homogeneous chemical reacti, J. Electroanal. Chem. 9 (5-6) (1965) 395–399. doi:10.1016/0022-0728(65)85037-9.

[42] H. Baltruschat, Differential electrochemical mass spectrometry, J. Am. Soc. Mass Spectrom. 15 (12) (2004) 1693–1706. doi:10.1016/j.jasms.2004.09.011.

[43] A.-E.-A. A. Abd-El-Latif, C. Bondue, S. Ernst, M. Hegemann, J. Kaul, M. Khodayari, E. Mostafa, A. Stefanova, H. Baltruschat, Insights into electrochemical reactions by differential electrochemical mass spectrometry, TrAC Trends Anal. Chem. 70 (2015) 4–13. doi:10.1016/j.trac.2015.01.015.

[44] M. Hess, T. Sasaki, C. Villevieille, P. Novák, Combined operando X-ray diffraction-electrochemical impedance spectroscopy detecting solid solution reactions of LiFePO4 in batteries, Nat. Commun. 6 (May) (2015) 1–9. doi:10.1038/ncomms9169.

[45] S. Crouch, D. Skoog, F. Holler, Principles of Instrumental Analysis, International Edition, Brooks/Cole, 2006.

[46] O. Wolter, J. Heitbaum, Differential Electrochemical Mass Spectroscopy (DEMS) - a new method for the study of electrode processes, Berichte der Bunsengesellschaft für Phys. Chemie 88 (1) (1984) 2–6. doi:10.1002/bbpc.19840880103.

[47] M. Fujihira, T. Noguchi, A novel differential electrochemical mass spectrometer (DEMS) with a stationary gas-permeable electrode in a rotational flow produced by a rotating rod, J. Electroanal. Chem. 347 (1993) 457–463.

[48] S. Wasmus, E. Cattaneo, W. Vielstich, Reduction of carbon dioxide to methane and ethene - an on-line MS study with rotating electrodes, Electrochim. Acta 35 (4) (1990) 771–775. doi:10.1016/0013-4686(90)90014-Q.

[49] D. Tegtmeyer, J. Heitbaum, A. Heindrichs, Electrochemical on line mass spectrometry on a rotating electrode inlet system, Berichte der Bunsengesellschaft für Phys. Chemie 93 (2) (1989) 201–206. doi:10.1002/bbpc.19890930218.

[50] I. Treufeld, A. J. J. Jebaraj, J. Xu, D. Martins de Godoi, D. Scherson, Porous Teflon ring-solid disk electrode arrangement for differential mass spectrometry measurements in the presence of convective flow generated by a jet impinging electrode in the wall-jet configuration, Anal. Chem. 84 (12) (2012) 5175–9. doi:10.1021/ac300799k.

[51] T. Hartung, H. Baltruschat, Differential electrochemical mass spectrometry using smooth electrodes: adsorption and hydrogen/deuterium exchange reactions of benzene on platinum, Langmuir 6 (5) (1990) 953–957. doi:10.1021/la00095a012.

[52] T. Hartung, U. Schmiemann, I. Kamphausen, H. Baltruschat, Electrodesorption from single-crystal electrodes: analysis by differential electrochemical mass spectrometry, Anal. Chem. 63 (1) (1991) 44–48. doi:10.1021/ac00001a008.

[53] Z. Jusys, H. Massong, H. Baltruschat, A new approach for simultaneous DEMS and EQCM: Electro-oxidation of adsorbed CO on Pt and Pt-Ru, J. Electrochem. Soc. 146 (3) (1999) 1093–1098. doi:10.1149/1.1391726.

[54] H. Wang, T. Löffler, H. Baltruschat, Formation of intermediates during methanol oxidation: A quantitative DEMS study, J. Appl. Electrochem. 31 (7) (2001) 759–765. doi:10.1023/A:1017539411059.

[55] H. Wang, L. R. Alden, F. J. DiSalvo, H. D. Abruña, Methanol electrooxidation on PtRu bulk alloys and carbon-supported PtRu nanoparticle catalysts: a quantitative DEMS study, Langmuir 25 (13) (2009) 7725–35. doi:10.1021/la900305k.

[56] A.-E.-A. A. Abd-El-Latif, J. Xu, N. Bogolowski, P. Königshoven, H. Baltruschat, New Cell for DEMS Applicable to Different Electrode Sizes, Electrocatalysis 3 (1) (2011) 39–47. doi:10.1007/s12678-011-0074-x.

[57] H. Wang, L. Alden, F. J. Disalvo, H. D. Abruña, Electrocatalytic mechanism and kinetics of SOMs oxidation on ordered PtPb and PtBi intermetallic compounds: DEMS and FTIRS study, Phys. Chem. Chem. Phys. 10 (25) (2008) 3739–51. doi:10.1039/b801473f.

[58] H. Wang, E. Rus, H. D. Abruña, New double-band-electrode channel flow differential electrochemical mass spectrometry cell: application for detecting product formation during methanol electrooxidation, Anal. Chem. 82 (11) (2010) 4319–24. doi:10.1021/ac100320a.

[59] A. Wonders, T. Housmans, V. Rosca, M. Koper, On-line mass spectrometry system for measurements at single-crystal electrodes in hanging meniscus configuration, J. Appl. Electrochem. 36 (11) (2006) 1215–1221. doi:10.1007/s10800-006-9173-4.

[60] Y. Gao, H. Tsuji, H. Hattori, H. Kita, New on-line mass spectrometer system designed for platinum-single crystal electrode and electroreduction of acetylene, J. Electroanal. Chem. 372 (1-2) (1994) 195–200. doi:10.1016/0022-0728(93)03291-V.

[61] K. Jambunathan, A. C. Hillier, Measuring electrocatalytic activity on a local scale with scanning Differential Electrochemical Mass Spectrometry, J. Electrochem. Soc. 150 (6) (2003) E312. doi:10.1149/1.1570823.

[62] S. Pérez-Rodríguez, M. Corengia, G. García, C. F. Zinola, M. J. Lázaro, E. Pastor, Gas diffusion electrodes for methanol electrooxidation studied by a new DEMS configuration: Influence of the diffusion layer, Int. J. Hydrogen Energy 37 (8) (2012) 7141–7151. doi:10.1016/j.ijhydene.2011.11.090.

[63] J. Flórez-Montaño, G. García, J. L. Rodríguez, E. Pastor, P. Cappellari, G. A. Planes, On the design of Pt based catalysts. Combining porous architecture with surface modification by Sn for electrocatalytic activity enhancement, J. Power Sources 282 (2015) 34–44. doi:10.1016/j.jpowsour.2015.02.018.

[64] C. Niether, M. Rau, C. Cremers, D. Jones, K. Pinkwart, J. Tübke, Development of a novel experimental DEMS set-up for electrocatalyst characterization under working conditions of high temperature polymer electrolyte fuel cells, J. Electroanal. Chem. 747 (2015) 97–103. doi:10.1016/j.jelechem.2015.04.002.

[65] T. Seiler, E. Savinova, K. A. Friedrich, U. Stimming, Poisoning of PtRu/C catalysts in the anode of a direct methanol fuel cell: a DEMS study, Electrochim. Acta 49 (22-23) (2004) 3927–3936. doi:10.1016/j.electacta.2004.01.081.

[66] N. Anastasijevic, H. Baltruschat, J. Heitbaum, DEMS as a tool for the investigation of dynamic processes: galvanostatic formic acid oxidation on a Pt electrode, J. Electroanal. Chem. Interfacial Electrochem. 272 (1-2) (1989) 89–100. doi:10.1016/0022-0728(89)87071-8.

[67] R. Schwiedernoch, S. Tischer, C. Correa, O. Deutschmann, Experimental and numerical study on the transient behavior of partial oxidation of methane in a catalytic monolith, Chem. Eng. Sci. 58 (3-6) (2003) 633–642. doi:10.1016/S0009-2509(02)00589-4.

[68] D. Zhang, O. Deutschmann, Y. E. Seidel, R. J. Behm, Interaction of mass transport and reaction kinetics during electrocatalytic co oxidation in a thin-layer flow cell, J. Phys. Chem. C 115 (2) (2011) 468–478. doi:10.1021/jp106967s.

[69] G. A. Planes, G. García, E. Pastor, High performance mesoporous Pt electrode for methanol electrooxidation. A DEMS study, Electrochem. commun. 9 (4) (2007) 839–844. doi:10.1016/j.elecom.2006.11.020.

[70] J. Fuhrmann, A. Linke, H. Langmach, H. Baltruschat, Numerical calculation of the limiting current for a cylindrical thin layer flow cell, Electrochim. Acta 55 (2) (2009) 430–438. doi:10.1016/j.electacta.2009.03.065.

[71] T. Chuah, J. Gimbun, T. S. Choong, A CFD study of the effect of cone dimensions on sampling aerocyclones performance and hydrodynamics, Powder Technol. 162 (2) (2006) 126–132. doi:10.1016/j.powtec.2005.12.010.

[72] T. Vidaković, M. Christov, K. Sundmacher, Investigation of electrochemical oxidation of methanol in a cyclone flow cell, Electrochim. Acta 49 (13) (2004) 2179–2187. doi:10.1016/j.electacta.2003.12.047.

[73] H. Uchida, K. Izumi, M. Watanabe, Temperature Dependence of CO-Tolerant Hydrogen Oxidation Reaction Activity at Pt, Pt-Co, and Pt-Ru Electrodes, J. Phys. Chem. B 110 (43) (2006) 21924–21930. doi:10.1021/jp064190x.

[74] M. Heinen, Y. X. Chen, Z. Jusys, R. J. Behm, Room temperature CO\superscript{ad} desorption/exchange kinetics on Pt electrodes - A combined in situ ir and mass spectrometry study, ChemPhysChem 8 (17) (2007) 2484–2489. doi:10.1002/cphc.200700425.

[75] T. Iwasita, U. Vogel, Interaction of methanol and CO adsorbate on platinum with CH3OH and CO in solution, Electrochim. Acta 33 (4) (1988) 557–560. doi:10.1016/0013-4686(88)80177-4.

[76] B. Geng, J. Cai, S. X. Liu, P. Zhang, Z. Q. Tang, D. Chen, Q. Tao, Y. X. Chen, S. Z. Zou, Temperature programmed Desorption - An application to kinetic studies of CO desorption at electrochemical interfaces, J. Phys. Chem. C 113 (47) (2009) 20152–20155. doi:10.1021/jp908481y.

[77] E. L. Cussler, Diffusion: Mass transfer in fluid systems, 2nd Edition, Cambridge University Press, New York, 1997.

[78] K. Sundmacher, Cyclone Flow cell for the investigation of gas-diffusion electrodes, J. Appl. Electrochem. 29 (1999) 919–926.

[79] U. T. Bödewadt, Die Drehströmung über festem Grunde, ZAMM - Zeitschrift für Angew. Math. und Mech. 20 (5) (1940) 241–253. doi:10.1002/zamm.19400200502.

[80] J. E. Nydahl, Heat Transfer for the Bödewadt Problem, Tech. rep., Colorado State University, Fort Collins, CO. (1971).

[81] W. J. Albery, S. Bruckenstein, Uniformly accessible electrodes, J. Electroanal. Chem. Interfacial Electrochem. 144 (1-2) (1983) 105–112. doi:10.1016/S0022-0728(83)80148-X.

[82] M. Bergelin, M. Wasberg, The impinging jet flow method in interfacial electrochemistry: an application to bead-type electrodes, J. Electroanal. Chem. 449 (1-2) (1998) 181–191. doi:10.1016/S0022-0728(98)00046-1.

[83] M. Bergelin, E. Herrero, J. M. Feliu, M. Wasberg, Oxidation of CO adlayers on Pt(111) at low potentials: an impinging jet study in H2SO4 electrolyte with mathematical modeling of the current transients, J. Electroanal. Chem. 467 (1) (1999) 74–84. doi:10.1016/S0022-0728(99)00046-7.

[84] J. J. Carroll, J. D. Slupsky, A. E. Mather, The solubility of carbon dioxide in water at low pressure, J. Phys. Chem. Ref. Data 20 (6) (1991) 1201. doi:10.1063/1.555900.

[85] C. Bondue, A.-E.-A. A. Abd-El-Latif, P. Hegemann, H. Baltruschat, Quantitative Study for Oxygen Reduction and Evolution in Aprotic Organic Electrolytes at Gas Diffusion Electrodes by DEMS, J. Electrochem. Soc. 162 (3) (2015) A479–A487. doi:10.1149/2.0871503jes.

[86] M. Khodayari, P. Reinsberg, A.-E.-A. A. Abd-El-Latif, C. Merdon, J. Fuhrmann, H. Baltruschat, Determining Solubility and Diffusivity by Using a Flow Cell Coupled to a Mass Spectrometer, ChemPhysChem 17 (11) (2016) 1647–1655. doi:10.1002/cphc.201600005.

[87] T. Vidaković, M. Christov, K. Sundmacher, A method for rough estimation of the catalyst surface area in a fuel cell, J. Appl. Electrochem. 39 (2) (2009) 213–225. doi:10.1007/s10800-008-9657-5.

[88] M. S. Masdar, T. Tsujiguchi, N. Nakagawa, Mass spectroscopy for the anode gas layer in a semi-passive DMFC using porous carbon plate Part I: Relationship between the gas composition and the current density, J. Power Sources 194 (2) (2009) 610–617. doi:10.1016/j.jpowsour.2009.07.027.

[89] J. Willsau, J. Heitbaum, The influence of Pt-activation on the corrosion of carbon in gas diffusion electrodes - A DEMS study, J. Electroanal. Chem. Interfacial Electrochem. 161 (1) (1984) 93–101. doi:10.1016/S0022-0728(84)80252-1.

[90] Mass Spectroscopy Soc. of Japan, MassBank Record: JP001576 (2011).

[91] P. J. Linstrom, W. Mallard (Eds.), NIST Chemistry WebBook, NIST Standard Reference Database Number 69, National Institute of Standards and Technology, Gaithersburg, 2016.

[92] E. Mostafa, A.-E.-A. A. Abd-El-Latif, H. Baltruschat, Electrocatalytic Oxidation and Adsorption Rate of Methanol at Pt Stepped Single-Crystal Electrodes and Effect of Ru Step Decoration: A DEMS Study, ChemPhysChem 15 (10) (2014) 2029–2043. doi:10.1002/cphc.201402193.

[93] S. Sakong, A. Groß, Methanol Oxidation on Pt(111) from First-Principles in Heterogeneous and Electrocatalysis, Electrocatalysis 8 (6) (2017) 577–586. doi:10.1007/s12678-017-0370-1.

[94] R. Babic, Y. Piljac, Kinetics and Electrocatalysis of Methanol Oxidation on Electrodeposited Pt and Pt 70 Ru 30 Catalysts, J. New Mater. Electrochem. Syst. 7 (2004) 179–190.

[95] T. Iwasita, Electrocatalysis of methanol oxidation, Electrochim. Acta 47 (22-23) (2002) 3663–3674. doi:10.1016/S0013-4686(02)00336-5.

[96] U. Krewer, H.-K. Yoon, H.-T. Kim, Basic model for membrane electrode assembly design for direct methanol fuel cells, J. Power Sources 175 (2) (2008) 760–772. doi:10.1016/j.jpowsour.2007.09.115.

[97] P. Hartmann, D. Grübl, H. Sommer, J. Janek, W. G. Bessler, P. Adelhelm, Pressure Dynamics in Metal–Oxygen (Metal–Air) Batteries: A Case Study on Sodium Superoxide Cells, J. Phys. Chem. C 118 (3) (2014) 1461–1471. arXiv:arXiv:1011.1669v3, doi:10.1021/jp4099478.

[98] D. Grübl, J. Janek, W. G. Bessler, Electrochemical Pressure Impedance Spectroscopy (EPIS) as Diagnostic Method for Electrochemical Cells with

Gaseous Reactants: A Model-Based Analysis, J. Electrochem. Soc. 163 (5) (2016) A599–A610. doi:10.1149/2.1041603jes.

[99] E. Engebretsen, T. J. Mason, P. R. Shearing, G. Hinds, D. J. Brett, Electrochemical pressure impedance spectroscopy applied to the study of polymer electrolyte fuel cells, Electrochem. commun. 75 (2017) 60–63. doi:10.1016/j.elecom.2016.12.014.

[100] A. Sorrentino, T. Vidakovic-Koch, K. Sundmacher, Studying mass transport dynamics in polymer electrolyte membrane fuel cells using concentration-alternating frequency response analysis, J. Power Sources 412 (November 2018) (2019) 331–335. doi:10.1016/j.jpowsour.2018.11.065.

[101] H. Wang, C. Wingender, H. Baltruschat, M. Lopez, M. Reetz, Methanol oxidation on Pt, PtRu, and colloidal Pt electrocatalysts: a DEMS study of product formation, J. Electroanal. Chem. 509 (2) (2001) 163–169. doi:10.1016/S0022-0728(01)00531-9.

[102] Z. Jusys, J. Kaiser, R. J. Behm, Composition and activity of high surface area PtRu catalysts towards adsorbed CO and methanol electrooxidation - A DEMS study, Electrochim. Acta 47 (22-23) (2002) 3693–3706. doi:10.1016/S0013-4686(02)00339-0.

[103] S. Wasmus, J.-T. Wang, R. F. Savinell, Real-Time Mass Spectrometric Investigation of the Methanol Oxidation in a Direct Methanol Fuel Cell, J. Electrochem. Soc. 142 (11) (1995) 3825. doi:10.1149/1.2048420.

[104] D. Ye, Untersuchung der Methanoloxidation mittels dynamischer DEMS Messung, Student thesis, TU Braunschweig (2015).

[105] N. Wolff, N. Harting, M. Heinrich, F. Röder, U. Krewer, Nonlinear Frequency Response Analysis on Lithium-Ion Batteries: A Model-Based Assessment, Electrochim. Acta 260 (2018) 614–622. doi:10.1016/j.electacta.2017.12.097.

[106] B. Yuan, Modellierung der Methanoloxidation auf Pt/Ru Katalysatoren in porösen Elektroden, Master thesis, TU Braunschweig (2015).

[107] F. Harnisch, K. Rabaey, The Diversity of Techniques to Study Electrochemically Active Biofilms Highlights the Need for Standardization, ChemSusChem 5 (6) (2012) 1027–1038. doi:10.1002/cssc.201100817.

[108] N. D. Rose, J. M. Regan, Changes in phosphorylation of adenosine phosphate and redox state of nicotinamide-adenine dinucleotide (phosphate) in Geobacter sulfurreducens in response to electron acceptor and anode potential variation, Bioelectrochemistry 106 (2015) 213–220. doi:10.1016/j.bioelechem.2015.03.003.

[109] P. A. Selembo, J. M. Perez, W. A. Lloyd, B. E. Logan, High hydrogen production from glycerol or glucose by electrohydrogenesis using microbial electrolysis cells, Int. J. Hydrogen Energy 34 (13) (2009) 5373–5381. doi:10.1016/j.ijhydene.2009.05.002.

[110] N. Montpart, L. Rago, J. A. Baeza, A. Guisasola, Hydrogen production in single chamber microbial electrolysis cells with different complex substrates, Water Res. 68 (2015) 601–615. doi:10.1016/j.watres.2014.10.026.

[111] J. Yu, Y. Park, T. Lee, Electron flux and microbial community in microbial fuel cells (open-circuit and closed-circuit modes) and fermentation, J. Ind. Microbiol. Biotechnol. 42 (7) (2015) 979–983. doi:10.1007/s10295-015-1629-2.

[112] A. Z. Andersen, F. R. Lauritsen, L. F. Olsen, On-line monitoring of CO2 production inLactococcus lactis during physiological pH decrease using membrane inlet mass spectrometry with dynamic pH calibration, Biotechnol. Bioeng. 92 (6) (2005) 740–747. doi:10.1002/bit.20641.

[113] M. J. Hayward, T. Kotiaho, A. K. Lister, R. G. Cooks, G. D. Austin, R. Narayan, G. T. Tsao, On-line monitoring of bioreactions of Bacillus polymyxa and Klebsiella oxytoca by membrane introduction tandem mass spectrometry with flow injection analysis sampling, Anal. Chem. 62 (17) (1990) 1798–1804. doi:10.1021/ac00216a014.

[114] F. R. Lauritsen, L. T. Nielsen, H. Degn, D. Lloyd, S. Bohátka, Identification of dissolved volatile metabolites in microbial cultures by membrane inlet tandem mass spectrometry, Biol. Mass Spectrom. 20 (5) (1991) 253–258. doi:10.1002/bms.1200200504.

[115] J. Das, H. Timm, H.-G. Busse, H. Degn, Oscillatory CO2 evolution in glycolysing yeast extracts, Yeast 6 (3) (1990) 255–261. doi:10.1002/yea.320060310.

[116] F. Kubannek, U. Krewer, A Cyclone Flow Cell for Quantitative Analysis of Kinetics at Porous Electrodes by Differential Electrochemical Mass Spectrometry, Electrochim. Acta 210 (2016) 862–873. doi:10.1016/j.electacta.2016.05.212.

[117] I. Schmidt, A. Pieper, H. Wichmann, B. Bunk, K. Huber, J. Overmann, P. J. Walla, U. Schröder, In Situ Autofluorescence Spectroelectrochemistry for the Study of Microbial Extracellular Electron Transfer, ChemElectroChem 4 (10) (2017) 2515–2519. doi:10.1002/celc.201700675.

[118] C. I. Torres, A. K. Marcus, P. Parameswaran, B. E. Rittmann, Kinetic Experiments for Evaluating the Nernst-Monod Model for Anode-Respiring Bacteria (ARB) in a Biofilm Anode, Environ. Sci. Technol. 42 (17) (2008) 6593–6597. doi:10.1021/es800970w.

[119] D. Millo, F. Harnisch, S. A. Patil, H. K. Ly, U. Schröder, P. Hildebrandt, In Situ Spectroelectrochemical Investigation of Electrocatalytic Microbial Biofilms by Surface-Enhanced Resonance Raman Spectroscopy, Angew. Chemie Int. Ed. 50 (11) (2011) 2625–2627. doi:10.1002/anie.201006046.

[120] J. R. Kim, B. Min, B. E. Logan, Evaluation of procedures to acclimate a microbial fuel cell for electricity production, Appl. Microbiol. Biotechnol. 68 (1) (2005) 23–30. doi:10.1007/s00253-004-1845-6.

[121] W. E. Balch, G. E. Fox, L. J. Magrum, C. R. Woese, R. S. Wolfe, Methanogens: reevaluation of a unique biological group, Microbiol. Rev. 43 (2) (1979) 260–296. arXiv:3697, doi:10.1016/j.watres.2010.10.010.

[122] N. S. Malvankar, T. Mester, M. T. Tuominen, D. R. Lovley, Supercapacitors Based on c-Type Cytochromes Using Conductive Nanostructured Networks of Living Bacteria, ChemPhysChem 13 (2) (2012) 463–468. doi:10.1002/cphc.201100865.

[123] R. Kumar, L. Singh, A. Zularisam, Exoelectrogens: Recent advances in molecular drivers involved in extracellular electron transfer and strategies used to improve it for microbial fuel cell applications, Renew. Sustain. Energy Rev. 56 (2016) 1322–1336. doi:10.1016/j.rser.2015.12.029.

[124] C. E. Levar, C. L. Hoffman, A. J. Dunshee, B. M. Toner, D. R. Bond, Redox potential as a master variable controlling pathways of metal reduction by Geobacter sulfurreducens, ISME J. 11 (3) (2017) 741–752. doi:10.1038/ismej.2016.146.

[125] A. S. Commault, G. Lear, M. A. Packer, R. J. Weld, Influence of anode potentials on selection of Geobacter strains in microbial electrolysis cells, Bioresour. Technol. 139 (2013) 226–234. doi:10.1016/j.biortech.2013.04.047.

[126] L. Zacharoff, C. H. Chan, D. R. Bond, Reduction of low potential electron acceptors requires the CbcL inner membrane cytochrome of Geobacter sulfurreducens, Bioelectrochemistry 107 (2016) 7–13. doi:10.1016/j.bioelechem.2015.08.003.

[127] Y. Liu, Y.-M. Lin, S.-F. Yang, A Thermodynamic Interpretation of the Monod Equation, Curr. Microbiol. 46 (3) (2003) 233–234. doi:10.1007/s00284-002-3934-z.

[128] Z. He, L. T. Angenent, Application of Bacterial Biocathodes in Microbial Fuel Cells, Electroanalysis 18 (19-20) (2006) 2009–2015. doi:10.1002/elan.200603628.

[129] L. Peng, X.-T. Zhang, J. Yin, S.-Y. Xu, Y. Zhang, D.-T. Xie, Z.-L. Li, Geobacter sulfurreducens adapts to low electrode potential for

extracellular electron transfer, Electrochim. Acta 191 (2016) 743–749. doi:10.1016/j.electacta.2016.01.033.

[130] H.-S. Lee, B. E. Rittmann, Significance of Biological Hydrogen Oxidation in a Continuous Single-Chamber Microbial Electrolysis Cell, Environ. Sci. Technol. 44 (3) (2010) 948–954. doi:10.1021/es9025358.

[131] S. Freguia, K. Rabaey, Z. Yuan, J. Keller, Electron and Carbon Balances in Microbial Fuel Cells Reveal Temporary Bacterial Storage Behavior During Electricity Generation, Environ. Sci. Technol. 41 (8) (2007) 2915–2921. doi:10.1021/es062611i.

[132] J. Song, D. Sasaki, K. Sasaki, S. Kato, A. Kondo, K. Hashimoto, S. Nakanishi, Comprehensive metabolomic analyses of anode-respiring Geobacter sulfurreducens cells: The impact of anode-respiration activity on intracellular metabolite levels, Process Biochem. 51 (1) (2016) 34–38. doi:10.1016/j.procbio.2015.11.012.

[133] V. B. Wang, K. Sivakumar, L. Yang, Q. Zhang, S. Kjelleberg, S. C. J. Loo, B. Cao, Metabolite-enabled mutualistic interaction between Shewanella oneidensis and Escherichia coli in a co-culture using an electrode as electron acceptor, Sci. Rep. 5 (1) (2015) 11222. doi:10.1038/srep11222.

[134] R. Moscoviz, F. de Fouchécour, G. Santa-Catalina, N. Bernet, E. Trably, Cooperative growth of Geobacter sulfurreducens and Clostridium pasteurianum with subsequent metabolic shift in glycerol fermentation, Sci. Rep. 7 (January) (2017) 44334. doi:10.1038/srep44334.

[135] C. Kim, Y. E. Song, C. R. Lee, B.-H. Jeon, J. R. Kim, Glycerol-fed microbial fuel cell with a co-culture of Shewanella oneidensis MR-1 and Klebsiella pneumonae J2B, J. Ind. Microbiol. Biotechnol. 43 (10) (2016) 1397–1403. doi:10.1007/s10295-016-1807-x.

[136] H. Kalfas, I. Skiadas, H. Gavala, K. Stamatelatou, G. Lyberatos, Application of ADM1 for the simulation of anaerobic digestion of olive pulp under mesophilic and thermophilic conditions, Water Sci. Technol. 54 (4) (2006) 149. doi:10.2166/wst.2006.536.

[137] M. Mahmoud, P. Parameswaran, C. I. Torres, B. E. Rittmann, Relieving the fermentation inhibition enables high electron recovery from landfill leachate in a microbial electrolysis cell, RSC Adv. 6 (8) (2016) 6658–6664. doi:10.1039/C5RA25918E.

[138] P. Parameswaran, C. I. Torres, H.-S. Lee, R. Krajmalnik-Brown, B. E. Rittmann, Syntrophic interactions among anode respiring bacteria (ARB) and Non-ARB in a biofilm anode: electron balances, Biotechnol. Bioeng. 103 (3) (2009) 513–523. doi:10.1002/bit.22267.

[139] V. K. Garlapati, U. Shankar, A. Budhiraja, Bioconversion technologies of crude glycerol to value added industrial products, Biotechnol. Reports 9 (2016) 9–14. doi:10.1016/j.btre.2015.11.002.

[140] G. P. da Silva, M. Mack, J. Contiero, Glycerol: A promising and abundant carbon source for industrial microbiology, Biotechnol. Adv. 27 (1) (2009) 30–39. doi:10.1016/j.biotechadv.2008.07.006.

[141] J. Marchetti, V. Miguel, A. Errazu, Possible methods for biodiesel production, Renew. Sustain. Energy Rev. 11 (6) (2007) 1300–1311. doi:10.1016/j.rser.2005.08.006.

[142] V. R. Nimje, C.-Y. Chen, C.-C. Chen, H.-R. Chen, M.-J. Tseng, J.-S. Jean, Y.-F. Chang, Glycerol degradation in single-chamber microbial fuel cells, Bioresour. Technol. 102 (3) (2011) 2629–2634. doi:10.1016/j.biortech.2010.10.062.

[143] T. Chookaew, P. Prasertsan, Z. J. Ren, Two-stage conversion of crude glycerol to energy using dark fermentation linked with microbial fuel cell or microbial electrolysis cell, N. Biotechnol. 31 (2) (2014) 179–184. doi:10.1016/j.nbt.2013.12.004.

[144] Y. Feng, Q. Yang, X. Wang, Y. Liu, H. Lee, N. Ren, Treatment of biodiesel production wastes with simultaneous electricity generation using a single-chamber microbial fuel cell, Bioresour. Technol. 102 (1) (2011) 411–415. doi:10.1016/j.biortech.2010.05.059.

[145] A. Tremouli, T. Vlassis, G. Antonopoulou, G. Lyberatos, Anaerobic Degradation of Pure Glycerol for Electricity Generation using a MFC: The Effect of Substrate Concentration, Waste and Biomass Valorization 7 (6) (2016) 1339–1347. doi:10.1007/s12649-016-9498-0.

[146] R. Moscoviz, E. Trably, N. Bernet, Consistent 1,3-propanediol production from glycerol in mixed culture fermentation over a wide range of pH, Biotechnol. Biofuels 9 (1) (2016) 32. doi:10.1186/s13068-016-0447-8.

[147] M. Zhou, J. Chen, S. Freguia, K. Rabaey, J. Keller, Carbon and Electron Fluxes during the Electricity Driven 1,3-Propanediol Biosynthesis from Glycerol, Environ. Sci. Technol. 47 (19) (2013) 11199–11205. arXiv:arXiv:1011.1669v3, doi:10.1021/es402132r.

[148] R. Kumar, V. P. Venugopalan, Development of self-sustaining phototrophic granular biomass for bioremediation applications, Curr. Sci. 108 (9) (2015) 1653–1661. arXiv:bit.20858, doi:10.1007/s12010-012-9609-8.

[149] J. F. Chignell, H. Liu, Biohydrogen Production From Glycerol in Microbial Electrolysis Cells and Prospects for Energy Recovery from Biodiesel Wastes, Proc. ASME 2011 Int. Manuf. Sci. Eng. Conf. MSEC2011-5 (2011) 1–9.

[150] Y. Sharma, R. Parnas, B. Li, Bioenergy production from glycerol in hydrogen producing bioreactors (HPBs) and microbial fuel cells (MFCs), Int. J. Hydrogen Energy 36 (6) (2011) 3853–3861. doi:10.1016/j.ijhydene.2010.12.040.

[151] A. Escapa, M.-F. Manuel, A. Morán, X. Gómez, S. R. Guiot, B. Tartakovsky, Hydrogen Production from Glycerol in a Membraneless Microbial Electrolysis Cell, Energy & Fuels 23 (9) (2009) 4612–4618. doi:10.1021/ef900357y.

[152] F. Kubannek, C. Moß, K. Huber, J. Overmann, U. Schröder, U. Krewer, Concentration Pulse Method for the Investigation of Transformation Pathways in a Glycerol-Fed Bioelectrochemical System, Front. Energy Res. 6. doi:10.3389/fenrg.2018.00125.

[153] R. Moscoviz, E. Trably, N. Bernet, Electro-fermentation triggering population selection in mixed-culture glycerol fermentation, Microb. Biotechnol.doi:10.1111/1751-7915.12747.

[154] H. Biebl, K. Menzel, A.-P. Zeng, W.-D. Deckwer, Microbial production of 1,3-propanediol, Appl. Microbiol. Biotechnol. 52 (3) (1999) 289–297. doi:10.1007/s002530051523.

[155] S. S. Yazdani, R. Gonzalez, Anaerobic fermentation of glycerol: a path to economic viability for the biofuels industry, Curr. Opin. Biotechnol. 18 (3) (2007) 213–219. doi:10.1016/j.copbio.2007.05.002.

[156] S. Riedl, R. K. Brown, S. Klöckner, K. J. Huber, B. Bunk, J. Overmann, U. Schröder, Successive Conditioning in Complex Artificial Wastewater Increases the Performance of Electrochemically Active Biofilms Treating Real Wastewater, ChemElectroChem 4 (12) (2017) 3081–3090. doi:10.1002/celc.201700929.

[157] T. Lueders, M. Manefield, M. W. Friedrich, Enhanced sensitivity of DNA- and rRNA-based stable isotope probing by fractionation and quantitative analysis of isopycnic centrifugation gradients, Environ. Microbiol. 6 (1) (2003) 73–78. doi:10.1046/j.1462-2920.2003.00536.x.

[158] A. K. Bartram, M. D. J. Lynch, J. C. Stearns, G. Moreno-Hagelsieb, J. D. Neufeld, Generation of Multimillion-Sequence 16S rRNA Gene Libraries from Complex Microbial Communities by Assembling Paired-End Illumina Reads, Appl. Environ. Microbiol. 77 (11) (2011) 3846–3852. arXiv:arXiv:1011.1669v3, doi:10.1128/AEM.02772-10.

[159] E. Aronesty, Comparison of Sequencing Utility Programs, Open Bioinforma. J. 7 (1) (2013) 1–8. doi:10.2174/1875036201307010001.

[160] R. C. Edgar, B. J. Haas, J. C. Clemente, C. Quince, R. Knight, UCHIME improves sensitivity and speed of chimera detection, Bioinformatics 27 (16) (2011) 2194–2200. doi:10.1093/bioinformatics/btr381.

[161] J. R. Cole, Q. Wang, J. A. Fish, B. Chai, D. M. McGarrell, Y. Sun, C. T. Brown, A. Porras-Alfaro, C. R. Kuske, J. M. Tiedje, Ribosomal Database Project: data and tools for high throughput rRNA analysis, Nucleic Acids Res. 42 (D1) (2014) D633–D642. doi:10.1093/nar/gkt1244.

[162] Q. Wang, G. M. Garrity, J. M. Tiedje, J. R. Cole, Naive Bayesian Classifier for Rapid Assignment of rRNA Sequences into the New Bacterial Taxonomy, Appl. Environ. Microbiol. 73 (16) (2007) 5261–5267. arXiv:Wang, Qiong, 2007, Naive, doi:10.1128/AEM.00062-07.

[163] Y. Liu, D. Deng, X. Lan, A highly Efficient mixed-culture biofilm as anodic catalyst and insights into its enhancement through electrochemistry by comparison with G. sulfurreducens, Electrochim. Acta 155 (2015) 327–334. doi:10.1016/j.electacta.2014.12.152.

[164] F. Kubannek, U. Schröder, U. Krewer, Revealing metabolic storage processes in electrode respiring bacteria by differential electrochemical mass spectrometry, Bioelectrochemistry 121 (2018) 160–168. doi:10.1016/j.bioelechem.2018.01.014.

[165] P. D. Kiely, J. M. Regan, B. E. Logan, The electric picnic: synergistic requirements for exoelectrogenic microbial communities, Curr. Opin. Biotechnol. 22 (3) (2011) 378–385. doi:10.1016/j.copbio.2011.03.003.

[166] A. R. Hari, K. P. Katuri, B. E. Logan, P. E. Saikaly, Set anode potentials affect the electron fluxes and microbial community structure in propionate-fed microbial electrolysis cells, Sci. Rep. 6 (1) (2016) 38690. doi:10.1038/srep38690.

[167] K.-J. Chae, M.-J. Choi, J.-W. Lee, K.-Y. Kim, I. S. Kim, Effect of different substrates on the performance, bacterial diversity, and bacterial viability in microbial fuel cells, Bioresour. Technol. 100 (14) (2009) 3518–3525. doi:10.1016/j.biortech.2009.02.065.

[168] P. F. Pind, I. Angelidaki, B. K. Ahring, Dynamics of the anaerobic process: Effects of volatile fatty acids, Biotechnol. Bioeng. 82 (7) (2003) 791–801. doi:10.1002/bit.10628.

[169] A. Esteve-Nunez, M. Rothermich, M. Sharma, D. R. Lovley, Growth of Geobacter sulfurreducens under nutrient-limiting conditions in continuous culture, Environ. Microbiol. 7 (5) (2005) 641–648. doi:10.1111/j.1462-2920.2005.00731.x.

[170] M. D. Yates, P. D. Kiely, D. F. Call, H. Rismani-Yazdi, K. Bibby, J. Peccia, J. M. Regan, B. E. Logan, Convergent development of anodic bacterial communities in microbial fuel cells, ISME J. 6 (11) (2012) 2002–2013. doi:10.1038/ismej.2012.42.

[171] A. I. Qatibi, V. Nivière, J. L. Garcia, Desulfovibrio alcoholovorans sp. nov., a sulfate-reducing bacterium able to grow on glycerol, 1,2- and 1,3-propanediol, Arch. Microbiol. 155 (2) (1991) 143–148. doi:10.1007/BF00248608.

[172] A. Qatibi, J. Cayol, J. Garcia, Glycerol and propanediols degradation by Desulfovibrio alcoholovorans in pure culture in the presence of sulfate, or in syntrophic association with Methanospirillum hungatei, FEMS Microbiol. Lett. 85 (3) (1991) 233–240. doi:10.1111/j.1574-6968.1991.tb04729.x.

[173] F. A. Lopes, P. Morin, R. Oliveira, L. F. Melo, Interaction of Desulfovibrio desulfuricans biofilms with stainless steel surface and its impact on bacterial metabolism, J. Appl. Microbiol. 101 (5) (2006) 1087–1095. doi:10.1111/j.1365-2672.2006.03001.x.

[174] D. F. Dwyer, J. M. Tiedje, Metabolism of polyethylene glycol by two anaerobic bacteria, Desulfovibrio desulfuricans and a Bacteroides sp., Appl. Environ. Microbiol. 52 (4) (1986) 852–856.

[175] P. Parameswaran, C. I. Torres, H.-S. Lee, B. E. Rittmann, R. Krajmalnik-Brown, Hydrogen consumption in microbial electrochemical systems (MXCs): The role of homo-acetogenic bacteria, Bioresour. Technol. 102 (1) (2011) 263–271. doi:10.1016/j.biortech.2010.03.133.

[176] J. De Vrieze, S. Gildemyn, J. B. Arends, I. Vanwonterghem, K. Verbeken, N. Boon, W. Verstraete, G. W. Tyson, T. Hennebel, K. Rabaey, Biomass retention on electrodes rather than electrical current enhances stability in anaerobic digestion, Water Res. 54 (2014) 211–221. doi:10.1016/j.watres.2014.01.044.

[177] K. L. Lesnik, H. Liu, Establishing a core microbiome in acetate-fed microbial fuel cells, Appl. Microbiol. Biotechnol. 98 (9) (2014) 4187–4196. doi:10.1007/s00253-013-5502-9.

[178] T. C. Pannell, R. K. Goud, D. J. Schell, A. P. Borole, Effect of fed-batch vs. continuous mode of operation on microbial fuel cell performance treating biorefinery wastewater, Biochem. Eng. J. 116 (2016) 85–94. doi:10.1016/j.bej.2016.04.029.

[179] V. Ortiz-Martinez, M. Salar-Garcia, A. de los Rios, F. Hernandez-Fernandez, J. Egea, L. Lozano, Developments in microbial fuel cell modeling, Chem. Eng. J. 271 (2015) 50–60. doi:10.1016/j.cej.2015.02.076.

[180] R. Sedaqatvand, M. Nasr Esfahany, T. Behzad, M. Mohseni, M. M. Mardanpour, Parameter estimation and characterization of a single-chamber microbial fuel cell for dairy wastewater treatment, Bioresour. Technol. 146 (2013) 247–253. doi:10.1016/j.biortech.2013.07.054.

[181] Y. Zeng, Y. F. Choo, B.-H. Kim, P. Wu, Modelling and simulation of two-chamber microbial fuel cell, J. Power Sources 195 (1) (2010) 79–89. doi:10.1016/j.jpowsour.2009.06.101.

[182] A. Donoso-Bravo, J. Mailier, C. Martin, J. Rodríguez, C. A. Aceves-Lara, A. V. Wouwer, Model selection, identification and validation in anaerobic digestion: A review, Water Res. 45 (17) (2011) 5347–5364. doi:10.1016/j.watres.2011.08.059.

[183] D. J. Batstone, P. F. Pind, I. Angelidaki, Kinetics of thermophilic, anaerobic oxidation of straight and branched chain butyrate and valerate, Biotechnol. Bioeng. 84 (2) (2003) 195–204. doi:10.1002/bit.10753.

[184] D. Sun, J. Chen, H. Huang, W. Liu, Y. Ye, S. Cheng, The effect of biofilm thickness on electrochemical activity of Geobacter sulfurreducens, Int. J. Hydrogen Energy 41 (37) (2016) 16523–16528. doi:10.1016/j.ijhydene.2016.04.163.

[185] J. Heijnen, Bioenergetics of Microbial Growth, in: Encycl. Bioprocess Technol., John Wiley & Sons, Inc., Hoboken, NJ, USA, 2002, pp. 267–290. doi:10.1002/0471250589.ebt026.

[186] R. M. Lewis, A. Shepherd, V. Torczon, Implementing Generating Set Search Methods for Linearly Constrained Minimization, SIAM J. Sci. Comput. 29 (6) (2007) 2507–2530. doi:10.1137/050635432.

[187] J. D. Stigter, D. Joubert, J. Molenaar, Observability of Complex Systems: Finding the Gap, Sci. Rep. 7 (1) (2017) 16566. doi:10.1038/s41598-017-16682-x.

[188] J. A. Jacquez, T. Perry, Parameter estimation: local identifiability of parameters, Am. J. Physiol. Metab. 258 (4) (1990) E727–E736. doi:10.1152/ajpendo.1990.258.4.E727.

[189] Y. Bard, Nonlinear Parameter Estimation, Academic Press, New York, 1974.

[190] G. H. Golub, C. Reinsch, Singular value decomposition and least squares solutions, Numer. Math. 14 (5) (1970) 403–420. doi:10.1007/BF02163027.

[191] C. I. Torres, On the importance of identifying, characterizing, and predicting fundamental phenomena towards microbial electrochemistry applications, Curr. Opin. Biotechnol. 27 (2014) 107–114. doi:10.1016/j.copbio.2013.12.008.

[192] C. Picioreanu, I. M. Head, K. P. Katuri, M. C. van Loosdrecht, K. Scott, A computational model for biofilm-based microbial fuel cells, Water Res. 41 (13) (2007) 2921–2940. doi:10.1016/j.watres.2007.04.009.

[193] B. E. Rittman, The effect of shear stress on biofilm loss rate, Biotechnol. Bioeng. 24 (2) (1982) 501–506. doi:10.1002/bit.260240219.

[194] H. Liu, R. Ramnarayanan, B. E. Logan, Production of Electricity during Wastewater Treatment Using a Single Chamber Microbial Fuel Cell, Environ. Sci. Technol. 38 (7) (2004) 2281–2285. doi:10.1021/es034923g.

[195] P. Stapor, D. Weindl, B. Ballnus, S. Hug, C. Loos, A. Fiedler, S. Krause, S. Hroß, F. Fröhlich, J. Hasenauer, PESTO: Parameter EStimation TOolbox, Bioinformatics 34 (4) (2018) 705–707. doi:10.1093/bioinformatics/btx676.

[196] W. K. Hastings, Monte Carlo sampling methods using Markov chains and their applications, Biometrika 57 (1) (1970) 97–109. doi:10.1093/biomet/57.1.97.

[197] R. Schenkendorf, X. Xie, M. Rehbein, S. Scholl, U. Krewer, The Impact of Global Sensitivities and Design Measures in Model-Based Optimal Experimental Design, Processes 6 (4) (2018) 27. doi:10.3390/pr6040027.

[198] K. Hansen, S. Gylling, F. Lauritsen, Time- and concentration-dependent relative peak intensities observed in electron impact membrane inlet mass spectra, Int. J. Mass Spectrom. Ion Process. 152 (2-3) (1996) 143–155. doi:10.1016/0168-1176(95)04338-1.

Appendix A

Appendix

A.1 Additional experimental data for CO oxidation experiment

Figure A.1 shows the flow of CO and CO_2 into vacuum during the potential step experiments from chapter 3. The flow of CO was calculated from the background corrected MS signal at m/z = 28. A contribution of 18.4% from CO_2 to the signal at m/z = 28, which was determined during the MS calibration experiment, and a CO calibration constant of 0.099 C mol^{-1} are taken into account. The drop below zero right after the potential step is an artefact which results from the rapid rise in CO_2 production and the correction of the signal at m/z = 28 for the CO_2 contribution.

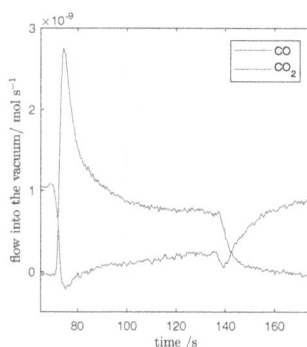

Figure A.1: Flow of CO and CO_2 into the vacuum during the potential step experiment.

A.2 Surface coverages for the cyclic voltammogram of the methanol oxidation reaction

In figure A.2, the simulated surface coverages of CO and OH during the CV are shown. The coverage of CO continuously decreases with potential. The OH coverage only increases at potentials above 0.65 V. The coverage of CO exhibits some hysteresis because in the positive scan, when the removal rate of CO from the surface is increasing, some time is needed before the surface coverage follows the potential. In the negative scan a delay occurs because the dissociative adsorption of methanol, that increases CO coverage, takes some time. The hysteresis is relatively small because of the low scan rate of $2\,\mathrm{mV\,s^{-1}}$.

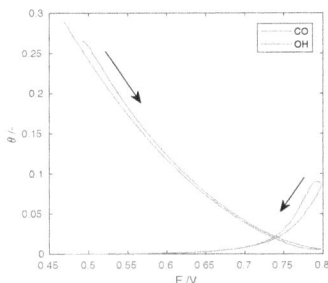

Figure A.2: Simulated surface coverages of CO and OH as a function of the potential for CV and MSCV, scan rate $2\,\mathrm{mV/s}$, $c_{\mathrm{MeOH}} = 0.5$ mol/L in $0.1\,\mathrm{mol/L}$ HClO$_4$, $0.5\,\mathrm{mg/cm^2}$ Pt/Ru on carbon, T=25° C.

A.3 Additional data and calculations for acetate oxidation in a biofilm electrode

A.3.1 Estimation of the ion current background

Here the determination of the ion current background for the DEMS experiments presented in chapter 5 is described.

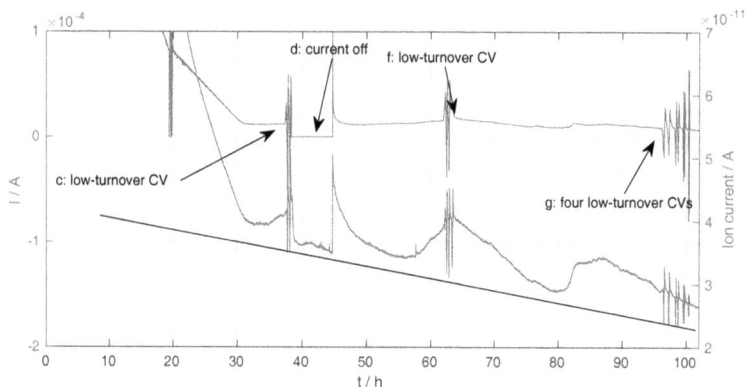

Figure A.3: Magnification of figure 5.2, showing the descending ion current base line. The minimal ion current during CV measurements (c, f, g) is dropping at approximately the same rate as the residual current during open circuit (d). The minimal ion current in the last two CVs is higher due to larger scan rates that do not allow the ion current to drop to the background level.

The background at $m/z = 44$ is slowly dropping throughout the experiment. Figure A.3 illustrates this almost linear trend. The fact that the background signal even persisted during six hours of operation under open circuit conditions (see arrow d in figure A.3) while the medium was continuously purged with nitrogen confirms that the background current was not related to a bioelectrochemical reaction. Therefore the background signal was substracted, which is common in DEMS measurements [85].

There is a number of possible reasons for the observed background signal. The background might shift because of imperfect purging of the re-circulation loop. Incomplete mixing or fluctuations in the gas flow of the purge gas might cause a variation in the degree of CO_2 removal. In general, it is difficult to remove CO_2 from a buffered neutral medium because it is mainly present in the form of carbonate [112]. Also, vacuum systems typically need a long time to reach steady state when the walls are not baked out. Transients of vacuum systems have been discussed for instance in [198].

A.3.2 Calculation of equilibrium potentials of acetate for different concentrations

In chapter 5, equilibrium potentials of acetate oxidation are discussed. In this section the calculation of these potentials is described.

$$E^0 = E^{00} + \frac{RT}{zF} \cdot \ln\left(\frac{a_{H+}^8}{a_{Ac}}\right) \tag{A.1}$$

$$\approx -148\,\text{mV} + \frac{8.314\,\text{J/mol/K} \cdot 308.15\,\text{K}}{8 \cdot 96485\,\text{C/mol}} \cdot (8\ln(c_{H+}) - \ln(c_{Ac})) \tag{A.2}$$

E^{00} was calculated from [128]. Activities were approximated by concentrations because of high dilution.

Acetate concentration/µM	5	10	20	40	100	10^6
E^0 (vs. sat. Ag/AgCl) /mV	-436	-439	-441	-443	-446	-477

Table A.1: Equilibrium potential for acetate oxidation at pH = 7 at different acetate concentrations

A.3.3 Mass Spectrometric Cyclic Voltammograms at all scan rates

Below MSCVs of acetate oxidation in a biofilm electrode recorded at different scan rates are presented.

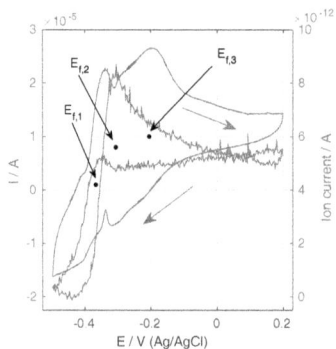

Figure A.4: MSCV at 0.5 mV/s. Ion current is plotted in orange, electrode current in blue. Black markers indicate the three formal potentials.

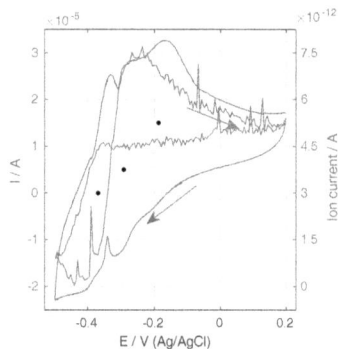

Figure A.5: MSCV at 1 mV/s. Ion current is plotted in orange, electrode current in blue. Black markers indicate the three formal potentials.

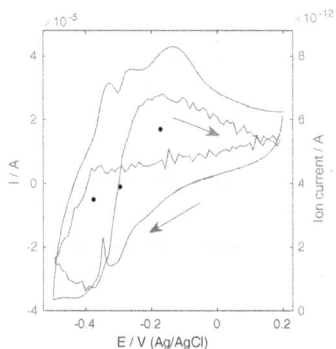

Figure A.6: MSCV at 2 mV/s. Ion current is plotted in orange, electrode current in blue. Black markers indicate the three formal potentials.

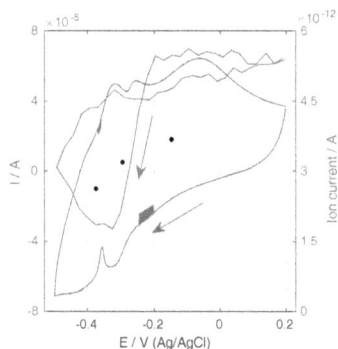

Figure A.7: MSCV at 5 mV/s. Ion current is plotted in orange, electrode current in blue. Black markers indicate the three formal potentials.

A.3.4 Mass Spectrometric Cyclic Voltammograms from the control experiment

Below MSCVs of acetate oxidation in a biofilm electrode recorded in a control experiment at different scan rates are presented. The overall shape of the graphs is the same as in the main experiment.

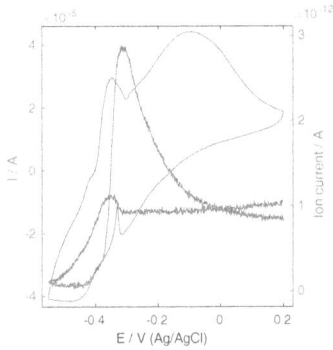

Figure A.8: MSCV at 0.2 mV/s. Ion current is plotted in orange, electrode current in blue.

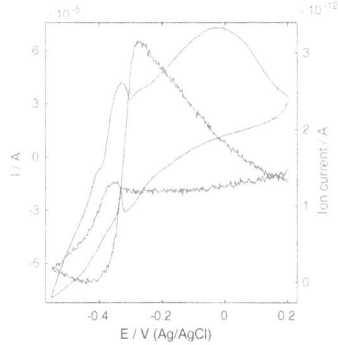

Figure A.9: MSCV at 0.5 mV/s. Ion current is plotted in orange, electrode current in blue.

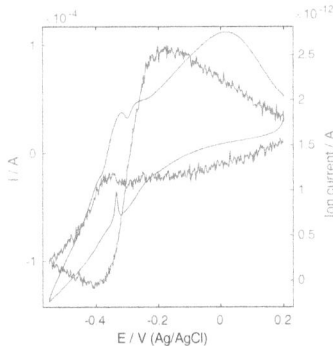

Figure A.10: MSCV at 1 mV/s. Ion current is plotted in orange, electrode current in blue.

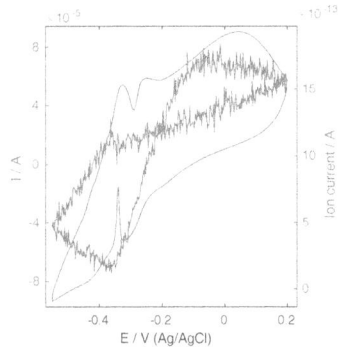

Figure A.11: MSCV at 2 mV/s. Ion current is plotted in orange, electrode current in blue.

A.3.5 Potential steps under turnover conditions

From the potential steps presented below the storage capacities of the biofilm electrode were calculated.

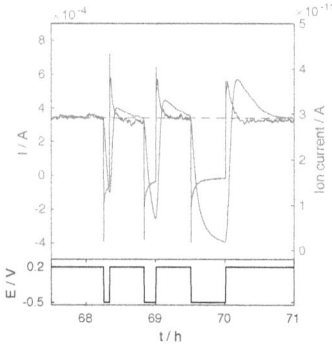

Figure A.12: Current and ion current upon stepping the potential from 0.2 to -0.5 V for 5, 10 and 30 minutes under turnover conditions. Ion current is plotted in orange, electrode current in blue.

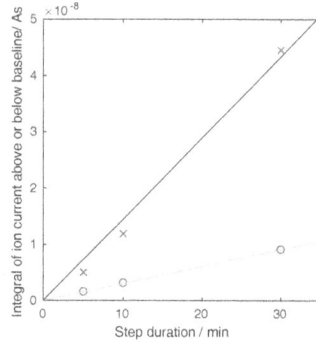

Figure A.13: Absolute values of the integrals of ion current below the baseline at -0.5,V and integrals of ion current above the baseline directly after stepping back to 0.2 V over the time for which the potential was set to -0.5 V under turnover conditions. Integrals above the baseline are plotted as circles, integrals below baseline as crosses. The continuous lines are linear regression functions.

A.4 Derivation of the effective half-saturation rate constant $K_{S,\text{eff}}$ for the biofilm model

Here the simplified relation to account for substrate mass transfer limitations within the biofilm, which is used in equation 6.9 in chapter 6.5, is derived.

Assuming that concentration gradients only occur in x-direction perpendicular to the electrode, the following substrate mass balance for the biofilm is established:

$$\frac{\mathrm{d}\,c}{\mathrm{d}\,t} = -D\frac{\partial^2 c}{\partial x^2} - r_\text{c} \tag{A.3}$$

c is the substrate concentration and D is the substrate diffusion coefficient within the biofilm. r_c is the volume-specific rate of substrate consumption. When the electrode potential is not limiting the substrate turnover rate, it can be described by Monod-Kinetics.

$$r_\text{c} = q_{\max}\rho_\text{bf}\frac{c}{c + K_\text{S}} \tag{A.4}$$

ρ_bf is the density of the biofilm biomass that is assumed to be constant here. $q_{\max}$ is the maximum specific substrate utilization rate and K_S is the half-saturation rate constant.

The following boundary conditions apply at the biofilm–electrode interface ($x = 0$) and at the biofilm–electrolyte interface ($x = \delta_\text{bf}$):

$$\left.\frac{\partial c}{\partial x}\right|_{x=0} = 0 \tag{A.5}$$

$$c\big|_{x=\delta_\text{bf}} = c_\text{bulk} \tag{A.6}$$

c_∞ is the substrate bulk concentration.

Solving the differential equation with the given boundary conditions allows to calculate the total rate of substrate consumption of the biofilm:

$$r_\text{total} = A_\text{el} \int_{x=0}^{\delta_\text{bf}} r_\text{c}\mathrm{d}x \tag{A.7}$$

Table A.2: Parameters for the demonstration of the transport model

Parameter	Value
A /cm^2	6
D /m^2 s^{-1}	$0.79 \cdot 10^{-9}$
ρ_{bf} /kg m^{-3}	150
c_{bulk} /mol m^{-3}	0.07
q_{max} /mol s^{-1} kg^{-1}	$4.2 \cdot 10^{-3}$
K_S /mol m^{-3}	0.03

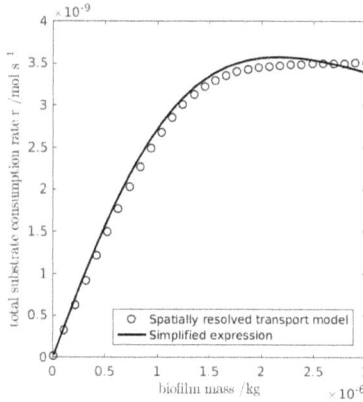

Figure A.14: Total rate of substrate consumption over the biomass of the biofilm.

The model was implemented and solved in Matlab with a finite volume scheme for biofilm thicknesses ranging from 0 to 50 µm, i.e. biofilm masses between 0 and 3 mg, using the parameters in Table A.2.

The rate of substrate consumption over the biofilm biomass is plotted in figure A.14 as open circles. The substrate consumption rate initially rises linearly with the biomass, flattens off, and finally approaches a maximum value. The initial linear dependency results from the fact that in a thin biofilm, where all layers are sufficiently supplied with substrate, all cells are metabolising at their specific maximum rate so that an increase in biomass causes an equal increase in substrate consumption rate. The upper limit of the substrate consumption rate is caused by the fact that at a certain biofilm thickness all substrate is consumed in the upper layers of the biofilm so that a further increase in biomass does not affect the total substrate consumption rate.

To simplify the model while still including the influence of transport resistances, an expression is developed that describes the total substrate consumption rate as a function of the total biofilm mass X_{bf} and the bulk concentration c_∞. This expression is chosen in order to resemble the shape of the Monod equation:

$$r_{\text{total}} = q_{\max} X_{\text{bf}} \frac{c_\infty}{c_\infty + K_{S,\text{eff}}} \tag{A.8}$$

$K_{S,\text{eff}}$ is the effective half saturation rate constant that includes the transport resistances in the film. Rearranging equation A.8 for $K_{S,\text{eff}}$ yields:

$$K_{S,\text{eff}} = \frac{q_{\max} X_{\text{bf}} c_\infty - r_{\text{total}} c_\infty}{r_{\text{total}}} \tag{A.9}$$

From the total rates of substrate consumption calculated with the spatially resolved model, $K_{S,\text{eff}}$ is calculated for various biofilm masses. In figure A.15 the values of $K_{S,\text{eff}}$ are plotted over the biofilm mass with open circles. If $K_{S,\text{eff}}$ would follow this curve exactly, the substrate consumption rate calculated by the simplified expression A.8 would equal that from the spatially resolved model.

In order to approximate the curve, the following equation is used, which is identical to equation 6.9 from chapter 6.5.

$$K_{S,\text{eff}} = K_{S,0} + K_{S,0} X_{\text{bf}}^2 \tag{A.10}$$

The values of $K_{S,\text{eff}}$ that were calculated with this simplified expression for $K_{S,0} = 0.03\,\text{mol}\,\text{m}^{-3}$ are plotted in figure A.14 with a solid line. The overall shape is reproduced well by the simplified expression with minor deviations at low and high biofilm masses. To evaluate the effect of these deviations, the total rates of substrate consumption were calculated according to equation A.8, substituting the values of $K_{S,\text{app}}$ by $K_{S,\text{eff}}$ calculated with equation A.10. In figure A.14, the rates of substrate consumption calculated from the spatially resolved biofilm model and from the simplified expression are compared. The two curves match very well for

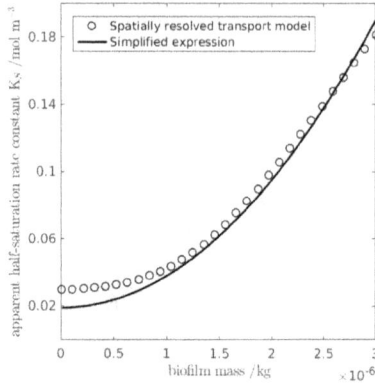

Figure A.15: Effective half-saturation rate constant $K_{S,\text{eff}}$ over biomass of the biofilm.

low values of biomass and still agree reasonably well for high values of biomass. The simplified expression A.10 is thus a suitable approximation for the transport processes inside the biofilm.

www.ingramcontent.com/pod-product-compliance
Lightning Source LLC
Chambersburg PA
CBHW060303220326
41598CB00027B/4221